Sven Bodo Wirsing

On the structure
of the Solomon-Tits algebra
of the symmetric group

An analysis of associative, group theoretic and Lie theoretical phenomenons

With 224 exercises

disserta
Verlag

Wirsing, Sven Bodo: On the structure of the Solomon-Tits algebra of the symmetric group. An analysis of associative, group theoretic and Lie theoretical phenomenons. With 224 exercises, Hamburg, disserta Verlag, 2022

Buch-ISBN: 978-3-95935-594-0
PDF-eBook-ISBN: 978-3-95935-595-7
Druck/Herstellung: disserta Verlag, Hamburg, 2022
Covermotiv: © designed by freepik.com

Bibliografische Information der Deutschen Nationalbibliothek:
Die Deutsche Nationalbibliothek verzeichnet diese Publikation in der Deutschen Nationalbibliografie; detaillierte bibliografische Daten sind im Internet über http://dnb.d-nb.de abrufbar.

© disserta Verlag, Imprint der Bedey & Thoms Media GmbH
Hermannstal 119k, 22119 Hamburg
http://www.disserta-verlag.de, Hamburg 2022
Printed in Germany

For Lena

Love 'Til Death

2

Contents

Introduction

Maybe we're all dreamers
I don't know where to go
Somewhere deep in the ocean (Ocean)
Where the wild fires glow
Lately got burned diving (Diving, diving)
I found a place where I can love my black heart
I used to think that you could fix me
I used to think that 'till death us do part
Are we not slaves?
Ghosts in the machine
Heaven is a waves
So clean and serene
Are we not slaves?
Slaves
Are we not? (Are we not, are we not, ...)
Give me the antidote (Antidote)
'Cause I really want to stay alive
I get my feeling that you're sinking (Sinking)
I'd like to pull you out in time, in time
Are we not slaves?
Ghosts in the machine
Heaven is a waves
So clean and serene
Are we not slaves?
Slaves
Slaves
Are we not?
So better keep on the side of you
So many people inside of you
Where the wind blows, where the wind blows, where the wind
    blows

8

Where the wind blows
I give you the best of me
Forty five, fifty five
I give you the best of me
Forty five, fifty five
Are we not slaves?
Slaves
Ghosts in the machine
Heaven is in waves
So clean and serene
Are we not slaves?
I wanted to be saved
Are we not slaves?
Slaves
Are we not?
I give you the best of me
Forty five, fifty five
Forty five, fifty five
(Lyrics Ghosts in the machine, Bush, 2020)

In the year 2003 Manfred Schocker gave a stimulating talk within the post-graduated seminar 'theory of algebras' (CAU, Kiel) about his research about the module structure of the Solomon-Tits algebra $\mathcal{T}_n$ of the symmetric group S_n. His results are released within the journal of algebra (see [18]; a preprint is available here: http://arxiv.org/abs/math/0505137). At this time I was a student at Kiel and participated at this post-graduated seminar led by Dieter Blessenohl and my doctoral advisor Hartmut Laue. The interesting lecture of Manfred Schocker was one of the motivations for studying the Solomon-Tits algebra.

The Solomon-Tits algebra $\mathcal{T}_n$ of S_n occur from a semigroup structure on the set of simplices of the Coxeter[1] complex $\sum_n$ associated with S_n and was first presented by Jacques Tits in an appendix to Solomon's original paper [20]. The simplices are in 1-1 correspondence with the ordered set partitions Π_n of the set $\underline{n} := \{1, \cdots, n\}$. The corresponding semigroup structure on Π_n

---

[1]Harold Scott MacDonald Coxeter

is defined as follows: Let $(P_1, \cdots, P_l)$ and $(Q_1, \cdots Q_k)$ elements of Π_n. Their product $\wedge_n$ is defined by

$$(P_1, \cdots, P_l) \wedge_n (Q_1, \cdots, Q_k) :=$$
$$(P_1 \cap Q_1, P_1 \cap Q_2, \cdots, P_1 \cap Q_k, \cdots, P_l \cap Q_1, P_l \cap Q_2, \cdots, P_l \cap Q_k)^{\emptyset}.$$

The symbol $^{\emptyset}$ indicates that empty sets are deleted within the tuple. The composition $\wedge_n$ defines a semigroup structure on Π_n with the two-sided identity $(\underline{n})$. Every element of Π_n is idempotent with respect to $\wedge_n$. If K is a field, then $\mathcal{T}_n$ is isomorphic to the monoid algebra $K\Pi_n$ which is of special interest within this work.

Manfred Schocker studies the module structure of the Solomon-Tits algebra of the symmetric group. His research includes among other things the construction of primitive idempotents, the decomposition into indecomposable principal modules (PIM), the description of the[2] Cartan-matrix, the determination of the nilindex of the nilradical, a description of a radical complement by using the primitive idempotents, the determination of the Ext-quiver and the description of the descending[3] Loewy-sequence. One of the key ideas of his research is the usage of a different basis which differs from the natural basis Π_n but consists still of idempotent elements e_P indexed by ordered set partitions. The results of Manfred Schocker's research build a basis for the analysis done within this work.

The symmetric group S_n acts on the set Π_n of ordered set partitions in a natural way on the components of the ordered set partitions by $(P_1, \cdots, P_l)\alpha :=$ $(P_1\alpha, \cdots, P_l\alpha)$ for all $(P_1, \cdots, P_l) \in \Pi_n$ and $\alpha \in S_n$. This action respects the product $\wedge_n$ on Π_n. Thus, the fixalgebra of S_n within $K\Pi_n$ is a subalgebra of the Solomon-Tits algebra. Patrick Bidigare proves within his article [6] that this fixalgebra is isomorphic to Solomons algebra D_n. Solomons algebra has a special role within the representation theory of the symmetric groups and is of special interests for several mathematicians within this area since many years. A very good overview about some main results are presented within

---

[2]Élie Cartan

[3]Alfred Loewy

10

the thesis of Manfred Schocker [18] and within the one of Thorsten Bauer [3], too.

The importance of Solomons algebra for several combinatorial and algebraical contexts is one cause for Thorsten Bauer to study within his thesis [3] the algebraical structure of Solomons algebra in characteristic zero. Again, one main insight for his study is the description of a second and suitable linear basis. Among other things he describes the stagnation of the ascending and descending Lie-central chains, its derivations and algebra automorphism, the Carter subgroups of its group of units and the Cartan subalgebras of its associated Lie algebra. The latter ones are studied in a more general context of finite-dimensional associative soluble algebras. Both structures are connected, too: the Carter subgroups are the group of units of the Cartan subalgebras. The research of Thorsten Bauer for Solomons algebra are an additional motivation for me to study the structure of the Solomon-Tits algebra.

The study within this work is done based on the following questions:

- Is it possible to generalize the results of Thorsten Bauer and to apply them to the Solomon-Tits algebra?

- Do more connections between the group of units and the Lie algebra[4] of an associative soluble algebra exist?

- What are insights about the associative structure, the group-theoretical structure of the group of units and the Lie algebra structure of the associated Lie algebra of Solomon-Tits algebra?

- Is it possible to apply the results to Solomons algebra?

These questions will not be answered completely within this work but they are the guidelines for our research. Now we present our results and the meaning of self-centralizing radical complements is highlighted as a more general context for studying our guidelines.

---

[4]Marius Sophus Lie

Chapter 1 has an introductory nature. Let K be a field. $K\Pi_n$ an associative idempotent monoid algebra, and this class of algebras is analyzed by Kenneth Brown in [7] using a special equivalence relation $\sim$ on an idempotent monoid M. Our focus is on the following topics:

- description of the nilradical of KM by using $\sim$

- splitting property of the factor algebra by its nilradical of KM and solvability of KM

- factorization of the monoid algebra KM in local components using $\sim$

- identification of the derivation of KM and $KM°$ as the nilradical of KM

- descriptions for the monoid algebra KM being commutative, separable, semisimple, simple or a division algebra

- examples of self-centralizing radical complements of the associative algebras $K\Pi_1$, $K\Pi_2$ and $K\Pi_3$.

Chapter 2 contains the following basic facts about $K\Pi_n$:

- construction of a second basis of $K\Pi_n$ which is one of the key ideas in Manfred Schocker's article [18]

- description of the nilradical and of one radical complement of $K\Pi_n$ by using this basis (see also [18])

- description of all idempotents of $K\Pi_n$

- determination of the K-span of all idempotents of $K\Pi_n$

- evaluation of the K-span of all separable elements for the cases being a radical complement or the whole algebra.

Within chapter 3 we focus on the following dimension-topics for $K\Pi_n$:

- dimension of $K\Pi_n$: counting ordered set partitions

- dimension of $(K\Pi_n)/(rad(K\Pi_n))$: counting unordered set partitions (Bell- and Stirling-numbers)

- dimension of $rad(K\Pi_n)$

- bounding these dimensions by using Solomons algebra

- growth of these dimensions.

Within chapter 4 we answer the following questions related to left and right ideals of $K\Pi_n$:

- Do special embeddings of $K\Pi_n$ in $K\Pi_{n+1}$ exists and what is the relation to right and left ideals?

- Is $K\Pi_n$ a (Quasi)-Frobenius algebra?

- Is $K\Pi_n$ uniserial?

- Is $K\Pi_n$ local?

- Is $K\Pi_n$ isomorphic to group algebra of a symmetric group?

- Does a special chain of ideals of $K\Pi_n$ exist which is based on the length-function?

- Do principal ideals in $K\Pi_n$ exist which are no principal right or left ideals (and the other combinations, too)?

- What is the connection of $K\Pi_n$ to Duo-algebras?

Within chapter 5 we focus on the following topics related to Duo algebras motivated by the previous chapter:

- answering remark 12 in the context of bi-modules

- consequences of the previous result

- characterizations of Duo algebras

- deduction of a necessary Lie property for Duo algebras and its application to $K\Pi_n$ and D_n.

Within chapter 6 we focus on the determination and description of Cartan subalgebras of the associated Lie algebra of $K\Pi_n$ and – in a more general context – of a finite-dimensional associative solvable K-algebra with splitting field K with respect to the following topics:

- connection between Pierce components and Cartan subalgebras

- description of self-centralizing radical components by special one-dimensiona Pierce components

- descriptions of the algebra, its nilradical and its radical complements based on Pierce components

- $(K\Pi_n)^\circ$ possesses only self-centralizing radical components

- determination and description of the Cartan subalgebras of $K\Pi_n$.

Within chapter 7 we focus on the (solvable) group of units of $K\Pi_n$ and – in a more general context – of a finite-dimensional associative solvable K-algebra A for a finite field of characteristic p with respect to the following topics::

- description of the Carter subgroups of $E(K\Pi_n)$

- conditions for $E(K\Pi_n)$ being nilpotent and commutative

- determination of the p-Sylow subgroup of $E(A)$

- determination and counting of the p'-Hall subgroups of $E(A)$ and its connection to the Carter subgroups

- applications to $E(K\Pi_n)$.

Within chapter 8 we focus on the center of $K\Pi_n$ and its group of units:

- determination of the center of $K\Pi_n$

- description of the center of $K\Pi_n$ by intersecting subalgebras

- conditions for $K\Pi_n$ being direct-indecomposable

- internal description of the center of Π_n by using classes with respect to $\sim$

- internal description of the center of the group of units of $K\Pi_n$

- external description of the center of the group of units of $K\Pi_n$ by using Carter subgroups

- connection between the center of the group of units and the group of units of the center of $K\Pi_n$.

Let A be a finite-dimensional associative unitary solvable K-algebra possessing a self-centralizing radical complement. Chapter 9 covers the following topics with respect to the stagnation of central chains of A° and $E(A)$:

- stagnation of the ascending central chain of A° at the center of A

- stagnation of the descending central chain of A° at the radical of A if A splits over K

- stagnation of the ascending central chain of $E(A)$ at the center of $E(A)$

- consequences of parts (i), (ii) and (iii) for $K\Pi_n$

- stagnation of the descending central chain of $E(K\Pi_n)$ and $E(D_n)$ at the derived subgroup by using commutator calculations with Pierce components and a sum-product-lemma that represents a vector-sum in $K\Pi_n$ resp. D_n as a product of units in $E(K\Pi_n)$ resp. $E(D_n)$.

Let K be a field and A a finite-dimensional associative unitary solvable K-algebra splitting over K such that a self-centralizing radical complement exists. Within chapter 10 we focus on the following topics with respect to classes of nilpotency and solvability:

- determining the Lie-product of k- and l-fold associative nilradical-powers

- determining the descending central chain of $rad(A)^\circ$

- calculating the class of nilpotency of $rad(A)^\circ$ by using the one of $rad(A)$

- determining the chain of derivations of $rad(A)^\circ$, A°, $rad(A)$ and A

- calculating the class of solvability of the previous mentioned structure by using the class of nilpotency of $rad(A)$

- deriving consequences for $K\Pi_n$ and D_n by using theorems of Manfred Schocker and M. D. Atkinson about the associative class of nilpotency of the nilradical

- calculating the commutator of k- and l-fold associative powers of $1 + rad(A)$ for $A = D_n$ and $A = K\Pi_n$ by using the sum-product-lemma

- determining the descending central chain of $1 + rad(A)$ for $A = D_n$ and $A = K\Pi_n$

- determining the class of nilpotency for $1 + rad(A)$ for $A = D_n$ and $A = K\Pi_n$ by using the class of nilpotency for $rad(A)$

- determining the chain of derivations for the group of units of $A = D_n$ and $A = K\Pi_n$ and for $1 + rad(A)$

- determining the class of solvability for the previous mentioned structures

- determining of exponents alongside the ascending and descending central chain and the derived chain within $1 + rad(K\Pi_n)$.

Within chapter 11 we focus on the connection of the Fitting subgroup and the Lie-nilradical motivated by a similar analysis done for Carter subgroups and Cartan subalgebras within the thesis of Thorsten Bauer in [3]:

- definition of a generalized Jordan decomposition and application to the adjoint representation

- determining the nilradical of the Lie algebra associated to a finite-dimensional associative unitary solvable algebra possessing a separable radical factor algebra

- determining the nilradical of D_n° and $K\Pi_n^\circ$

- determining the Fitting subgroup of the group of units of a finite-dimensional associative unitary solvable algebra possessing a separable radical factor algebra

- analyzing the connection of the Fitting subgroup and the Lie-nilradical

- determining the Fitting subgroup of $E(D_n)$ and $E(K\Pi_n)$.

Within chapter 12 we analyze semisimple left and right ideals (also for the determination of the anti-automorphism of D_n and $K\Pi_n$ in the next chapter). The following topics are of special interest for finite-dimensional associative solvable K-algebras splitting over K and possessing a self-centralizing radical complement:

- definition of left-sided, right-sided and two-sided Pierce-orthogonal elements

- description of maximal semisimple left and right ideals by using Pierce-orthogonal elements

- conjugation of theses maximal semisimple left and right ideals

- description of simple left and right ideals by using Pierce-orthogonal elements

- consequences for D_n and $K\Pi_n$.

Within chapter 13 we determine the anti-automorphism of D_n and $K\Pi_n$ and focus on the following topics:

- determination of the dimension of the maximal semisimple left and right ideals of D_n

- proof of the fact that almost no anti-automorphism of D_n exist

- determination of the anti-automorphism of D_n in the remaining cases of n

- determination of the maximal dimension of projective indecomposable left and right ideals of $K\Pi_n$

- proof of the fact that almost no anti-automorphism of $K\Pi_n$ exist

- determination of the anti-automorphism of $K\Pi_n$ in the remaining cases of n.

Within the last chapter we focus on the following topics related to irreducible characters of monoid algebras related to finite idempotent monoids:

- determination of irreducible characters of $K\Pi_n$ done already by Kenneth Brown

- description of the values 0 and 1 of the irreducible characters based on sub-semigroups

- determination of the number of elements of these sub-semigroups for Π_n

- consequences for the right ideals $P \cdot K\Pi_n$ and $e_P \cdot K\Pi_n$

- consequences for the left ideals $K\Pi_n \cdot P$ and $K\Pi_n \cdot e_P$.

$$* * *$$

Some applications are also transferred to the exercises at the end of each chapter. There are some exercises included enhancing the theory presented so far such that the reader can experience a deeper insight. In addition, at the beginning of each exercise series some open-ended topics are included which can be used by the reader – and also by the author – to do additional researches within the presented topic. The author has included some manually created graphics – mostly so called Hasse diagrams – to visualize the main results of this work.

Exercise 1 *Analyse and describe the connection between the guidelines of this work and the results presented within the introduction.*

Chapter 1

Idempotent monoid algebras and $K\Pi_n$

This chapter has an introductory nature. Let K be a field. $K\Pi_n$ an associative idempotent monoid algebra, and this class of algebras is analyzed by Kenneth Brown in [7] using a special equivalence relation $\sim$ on an idempotent monoid M. Our focus is on the following topics:

- description of the nilradical of KM by using $\sim$

- splitting property of the factor algebra by its nilradical of KM and solvability of KM

- factorization of the monoid algebra KM in local components using $\sim$

- identification of the derivation of KM and KM° as the nilradical of KM

- descriptions for the monoid algebra KM being commutative, separable, semisimple, simple or a division algebra

- examples of self-centralizing radical complements of the associative algebras $K\Pi_1$, $K\Pi_2$ and $K\Pi_3$.

1.1 Derivation, nilradical and its factor algebra

Definitions 1 A semigroup is called idempotent if every of its elements is idempotent. For every semigroup H and every field K let KH be the semigroup algebra with respect to K and H. If K is a field and A an

17

associative unitary K-algebra, then let $min_{a,K}$ be the minimal polynomial of an algebraic element a of A within the polynomial algebra $K[t]$. If M is a set and $n \in \mathbb{N}$, then let M^n be the set of n-tuples over M. If T is a subalgebra of an associative algebra and $n \in \mathbb{N}$, then we denote by $T^{<n>}$ the K-span of all products with n factors of elements of T. If T is nilpotent, then $cl(T)$ denotes the class of nilpotency (also called nilindex) of T.$\diamond$

Lemma 1 *Let K be a field and L a finite commutative idempotent monoid. The associative algebras KL and $K^{|L|}$ are isomorphic.*

Proof: For all $l \in L$ the equation $l^2 = l$ is valid, and thus we deduce $min_{l,K} \mid t^2 - t = t(t-1)$. Consequently, all elements of $L(\subseteq KL)$ are diagonalizable. L is commutative, and by using Satz 5.5.1 in [22] we conclude that KL is diagonalizable and isomorphic to $K^{|L|}$.$\diamond$

Definitions 2 For every semigroup H we define a relation $\sim_H$ by

$$a \sim_H b :\Leftrightarrow a = aba \wedge b = bab$$

for all $a, b \in H$. In addition, for all $a, b \in H$ we set

$$a <_H b :\Leftrightarrow a = aba \text{ and } a >_H b :\Leftrightarrow b <_H a.$$

Let K be a field and V be a finite-dimensional K-space with basis B. We define

$$Aug_B(V) := \{v \mid \forall b \in B\, \exists k_b \in K : v = \sum_{b \in B} k_b b \wedge \sum_{b \in B} k_b = 0\}.$$

In the special case of a finite semigroup H we symbolize this space by $Aug(KH)$ instead of using $Aug_H(KH)$ and call $Aug(KH)$ the augmentation ideal of KH.

For every subset T of a K-space V we symbolize the K-span of T in V by $\langle T \rangle_K$. Let A, B be subspaces of V with trivial intersection. In this case we call their sum direct and symbolize the direct sum by $A \oplus B$. For a finite-dimensional K-space V we use $dim_K(V)$ for its K-dimension. If W is another K-space, U a subspace of V and α a linear function between V and W, then let $Kern\,\alpha$ be the kernel of α and $\alpha_{|U}$ be the restriction of α to U. The identity (function) of V is symbolized by id_V.

For every associative K-algebra A we can use the multiplication $a \circ b := ab - ba$ for all $a, b \in A$ to define a Lie-algebra A° which is called the associated Lie-algebra of A and which is symbolized by A°. For subsets S, T of A let $S \circ T := \langle s \circ t \mid s \in S, t \in T \rangle_K$ be the K-span of Lie-commutators of S and T.

If A is an associative K-algebra, then $rad(A)$ is the nilradical and A' the derivation or derived subalgebra of A (the smallest ideal of A generated by $A \circ A$). A radical complement is a subalgebra T of A such that $A = rad(A) \oplus T$ is valid. A is solvable or soluble if the derivation of A is contained in the nilradical of A. For every subset T of A let $C_A(T)$ be the centralizer of T in A and $Z(A) := C_A(A)$ the center of A. If A is unitary, then $E(A)$ is the group of units of A. If $a \in A, x \in E(A)$, then let x^{-1} be the inverse of x and $a^x := x^{-1}ax$ be the conjugate element of a by x.

Let R be a set and $\sim$ be an equivalence relation on R. For every $r \in R$ we denote the unique equivalence class containing r by $[r]_\sim$.

For all $n \in \mathbb{N}$ we define $\underline{n} := \{i \mid i \in \mathbb{N}, 1 \leq i \leq n\}$ and $\underline{n}_0 := \underline{n} \cup \{0\}$.

If T is a finite subset of $\mathbb{N}$, then we define $min(T)$ resp. $max(T)$ the minimum resp. maximum of T.

If G is a group containing the subset T, then let $\langle T \rangle_G$ be the smallest subgroup of G containing T. $\diamond$

Remark 1 *Let K be a field and V a finite-dimensional K-space with K-basis B. For all $b' \in B$ the identity $Aug_B(V) = \langle b' - b \mid b \in B, b \neq b' \rangle_K$ is valid. In particular, $Aug_B(V)$ is a subspace of codimension 1 of V.* $\diamond$

Lemma 2 *Let H be an idempotent semigroup and K be a field.*

(i) *$\sim_H$ is an equivalence relation on H. (Kenneth Brown)*

(ii) *Every equivalence class of $\sim_H$ is a sub-semigroup of H. In particular, H is a disjoint union of sub-semigroups of H. If H is a monoid, then $[1]_{\sim_H} = \{1\}$ is valid. Thus, only in the case $\mid H \mid = 1$ exactly one equivalence class exists.*

(iii) *The set L of all equivalence classes of $\sim_H$ is – with respect to the multiplication $[a]_{\sim_H} [b]_{\sim_H} := [ab]_{\sim_H}$ for all $a, b \in H$ – an idempotent commutative semigroup and is called the abelianization of H.[1] If H is a monoid, then L is a monoid, too. (Kenneth Brown)*

(iv) *The function $P_H : h \longrightarrow [h]_{\sim_H}$ is an algebra epimorphism between KH and KL (L abelianization of H). (Kenneth Brown)*

(v) *For all $h \in H$ and finite $[h]_{\sim_H}$ the identity $Kern\,(P_H)_{\mid [h]_{\sim_H}} = Aug(K[h]_{\sim_H})$ is valid.*

---

[1]Niels Henrik Abel

Proof: ad(i), (iii), (iv): see [7]

ad(ii): Let $x, y \in H$ and $x \sim_H y$ be valid. We calculate:

$$xy \sim_H x$$
$$\Longleftrightarrow \quad xy = (xy)x(xy) \quad \wedge \quad x = x(xy)x$$
$$\Longleftrightarrow \quad xy = xy(xx)y \quad \wedge \quad x = (xx)yx$$
$$\Longleftrightarrow \quad xy = (xy)(xy) \quad \wedge \quad x = xyx$$
$$\Longleftrightarrow \quad xy = xy \quad \wedge \quad x = xyx.$$

The condition $x = xyx$ is true based on $x \sim_H y$.
For all $h \in H$ the identity $h \sim_H 1$ is true if and only if $h = h1h$ and $1 = 1h1$ are valid. The latter ones are equivalent to $h = 1$.

ad(v): Let $x \in K[h]_{\sim_H}$ and $T := [h]_{\sim_H}$. Elements $k_t \in K$ exist for all $t \in T$ such that $x = \sum_{t \in T} k_t t$ is true. $xP_H = 0$ is equivalent to $(\sum_{t \in T} k_t)T = 0.\diamond$

Proposition 1 *Let H be an idempotent semigroup, K a field, $n \in \mathbb{N}$ and $x, y \in H$ such that $x \sim_H y$ is valid.*

(i) $(x - y)^3 = 0$
   For $H = \Pi_n$ the identity $(x - y)^2 = 0$ is true.

(ii) $y - x = x \circ (xy) - y \circ (yx) - x \circ y$

Proof: ad(i): By using $x \sim_H y$ we deduce $x = xyx$ and $y = yxy$. We calculate:

$$(x - y)^2$$
$$= (x - y)(x - y)$$
$$= xx - xy - yx + yy$$
$$= x - xy - yx + y, \text{ and}$$

$$(x - y)^3$$
$$= (x - xy - yx + y)(x - y)$$
$$= xx - xy - xyx + xyy - yxx + yxy + yx - yy$$
$$= x - xy - xyx + xy - yx + yxy + yx - y$$
$$= (x - xyx) + (yxy - y)$$
$$= 0 + 0.$$

Based on page 9 in [18] within Π_n the identity $xyx = xy$ is valid for all $x, y \in \Pi_n$.

ad(ii): Because of $x = xyx$ and $y = yxy$ we deduce:

$$
\begin{aligned}
& x \circ (xy) - y \circ (yx) - x \circ y \\
=~& (xy - xyx) - (yyx - yxy) - (xy - yx) \\
=~& (xy - x) - (yx - y) - (xy - yx) \\
=~& y - x. \diamond
\end{aligned}
$$

Theorem 1 *(local version) Let K be a field and H a finite idempotent semigroup such that H is the only equivalence class of $\sim_H$.*

(i) *$rad(KH) = Kern\, P_H = Aug(KH)$ (Kenneth Brown,[7])*

(ii) *$KH/rad(KH) \cong K$ (Kenneth Brown,[7])*
   In particular, $K \cdot 1$ is a radical complement.

(iii) *$rad(KH) = (KH)' = KH \circ KH$.*

Proof: ad(i),(ii): By using part (v) of lemma 2 the identity $Kern\, P_H = Aug(KH)$ is valid. We prove that $Kern\, P_H$ is nilpotent and that $KH/Kern\, P_H$ and K are isomorphic. Consequently, (i) and (ii) are valid. Because of remark 1 we deduce that $Kern\, P_H$ is K-linear generated by $\{x - y \mid x, y \in H, x \sim_H y\}$. This set consists – by using part (i) of proposition 1 – of nilpotent elements. A theorem[2] of Joseph Wedderburn (see e.g. [14]) is now used to prove that the ideal $Kern\, P_H$ is nilpotent. The factor algebra KH by $Kern\, P_H$ is – based on part (iii) of lemma 2 – isomorphic to KL (L the abelianization of H). Under our assumptions $L = \{H\}$ is valid. Lemma 1 and part (iii) of lemma 2 let us deduce that the factor algebra KH by $Kern\, P_H$ and K are isomorphic.

ad(iii): KH is solvable based on part (i) and its nilradical contains the derivation of KH. By definition, the derivation of KH and the ideal generated by $KH \circ KH$ are identical. We use (i) and remark 1 to conclude that we only have to prove now that for all $x, y \in H$ such that $x \sim_H y$ is valid the element $x - y$ is contained in $KH \circ KH$. But this statement is a direct

---

[2]Joseph Henry Maclagan Wedderburn

22

consequence of part (ii) of proposition 1.◇

Let K be a field and A an associative K-algebra. K is a splitting field for $A/rad(A)$ if and only if this algebra is isomorphic to a direct sum of full matrix algebras over K: $\bigoplus_{i=1}^{n} K^{n_i \times n_i}$. Sometimes also the statement is used that K is a splitting field for A although $A/rad(A)$ splits. In the solvable case of A the splitting is equivalent to $A/rad(A) \cong K^n$ for some $n \in \mathbb{N}$. Such algebras are also called diagonalizable. The first identity resp. the isomorphism in parts (vi) and (vii) of the next theorem were already proven by Kenneth Brown (see [7]):

Theorem 2 *(global version) Let K be a field, M a finite idempotent monoid and L the abelianization of M. The following statements are valid:*

(i) $KM = \bigoplus_{T \in L} KT$
   (direct sum of K-subalgebras)

(ii) $Kern\, P_M = \bigoplus_{T \in L} Kern\, P_T$

(iii) $rad(KM) = \bigoplus_{T \in L} rad(KT)$

(iv) $(KM)' = \bigoplus_{T \in L} (KT)'$

(v) $KM \circ KM = \bigoplus_{T \in L} (KT \circ KT)$

(vi) $KM/rad(KM) \cong K^{|L|} \cong \bigoplus_{T \in L} (KT/rad(KT))$
   In particular, radical complements exist and they are conjugated under $1 + rad(KM)$. KM is solvable and K is a splitting field for $KM/rad(KM)$.

(vii) $rad(KM) = Kern\, P_M = (KM)' = KM \circ KM$

(viii) $\bigoplus_{T \in L} Aug(KT) \subseteq Aug(KM)$
   The identity is only valid in the case $|L| = 1$.

Proof: ad(i): This statement is a direct consequence of parts (i) and (ii) of lemma 2.

ad(ii): Let $x \in KM$, about $x = \sum\limits_{T \in L} \sum\limits_{t \in T} k_{t,T} t$. By using part (i) and part (iv) of lemma 2 we deduce:

$$x \in Kern\, P_M$$
$$\iff \sum_{T \in L} \sum_{t \in T} k_{t,T} T = 0$$
$$\iff \forall T \in L : \sum_{t \in T} k_{t,T} = 0.$$

ad(iii)-(viii): By using part (ii) and theorem 1 we calculate:

$$Kern\, P_M$$
$$= \bigoplus_{T \in L} Kern\, P_T$$
$$= \bigoplus_{T \in L} rad(KT)$$
$$= \bigoplus_{T \in L} (KT)'$$
$$= \bigoplus_{T \in L} (KT \circ KT)$$
$$= \bigoplus_{T \in L} Aug(KT).$$

Because of $Kern\, P_M = \bigoplus\limits_{T \in L} rad(KT)$ this ideal is K-linear generated by nilpotent elements. Therefor, it is nilpotent by using a theorem of Joseph Wedderburn (see e.g. [14]). Part (iv) of lemma 2 and lemma 1 let us deduce that the factor algebra of KM by $Kern\, P_M$ is isomorphic to the algebra $K^{|L|}$. Hence, $rad(KM) = Kern\, P_M$ is valid. KM is solvable, and thus $KM \circ KM \subseteq (KM)' \subseteq rad(KM)$ is true. Because of $rad(KM) = \bigoplus\limits_{T \in L} (KT \circ KT)$ we deduce $rad(KM) = KM \circ KM$. In particular, parts (iii)-(vii) are valid. $Aug(KM)$ is – based on remark 1 – of codimension 1 in KM and contains $\bigoplus\limits_{T \in L} Aug(KT)$. This space is of (using the same remark) codimension $| L |$ in KM. The add-on in (vi) is true based on the theorem of Wedderburn-Malcev[3].◇

---

[3]Anatoli Iwanowitsch Malzew

24

Corollary 1 *Let K be a field, M a finite idempotent monoid and T a sub-algebra of KM. K is a slitting field for $T/rad(T)$ and T is solvable.*

Proof: see theorem 3.3.2 in [22] and theorem 2, part (vi).⋄

Corollary 2 *Let K be a field and $n \in \mathbb{N}$. K is a splitting field for Solomons algebra D_n and D_n is solvable.*

Proof: Π_n is a finite idempotent monoid, and thus $K\Pi_n$ is fulfilling the assumptions of theorem 2. In particular, every subalgebra of $K\Pi_n$ is – based on corollary 1 – splitting over K. Patrick Bidigare has proven within in [6] that an isomorphic copy of Solomons algebra D_n is contained in $K\Pi_n$. Hence, K is a splitting field for D_n.⋄

Corollary 3 *Let K be a field and M a finite idempotent monoid. KM is a local algebra if and only if M consists of a single element only.*

Proof: see part (vi) of theorem 2 and pert (ii) of lemma 2.⋄

Corollary 4 *Let K be a field and M a finite idempotent monoid. The following statements are equivalent:*

(i) M possesses only one element.

(ii) The associative algebras KM and K are isomorphic.

(iii) KM is a division algebra.

(iv) KM is simple.

Proof: see part (vi) of theorem 2 and part (ii) of lemma 2.⋄

Corollary 5 *Let K be a field and M a finite idempotent monoid. The nilradical of KM is K-linear generated by $\{x - y \mid x \sim_M y\}$.*

Proof: see remark 1, part (i) of theorem 1 and part (iii) of theorem 2.⋄

Corollary 6 *Let K be a field and M a finite idempotent monoid. The following statements are equivalent:*

(i) KM is separable.

(ii) KM is semisimple.

(iii) M is commutative.

(iv) KM and $K^{|M|}$ are isomorphic.

(v) Every equivalence class of M with respect to $\sim_M$ consists of a single element only.

Proof: Every separable algebra is semisimple. By using the semisimplicity and part (vi) of theorem 2 we deduce the commutativity of KM. If KM is commutative, then lemma 1 lets us deduce that KM and $K^{|M|}$ are isomorphic and henceforth they are separable. Therefore, the parts (i)-(iv) are equivalent. Based on corollary 5 we conclude that the nilradical of KM ist K-linear generated by the set $\{x - y \mid x \sim_M y\}$. As a consequence, parts (ii) and (v) are equivalent. $\diamond$

Remark 2 Let K be a field and $n \in \mathbb{N}$. We will prove later within this work that all statements of corollaries 3, 4 and 6 are equivalent for $M = \Pi_n$. $\diamond$

Remark 3 Let M be an idempotent monoid. Within corollary 6 we have proven for M being finite that M is commutative if and only if every class with respect to $\sim_M$ consists of one element only. We want to re-prove this result by calculation done purely within M. For every subset T of M let $C_M(T)$ be the centralizer of T in M and especially $Z(M) := C_M(M)$ be the center of M. Instead of $C_M(\{a_1, ..., a_n\})$ we will use the notation $C_M(a_1, ..., a_n)$ for arbitrary finite many $a_1, ..., a_n \in M$. Let $a, b \in M$. Now we will prove:

(i) $ab \sim_M ba$

(ii) $aba \sim_M ba$

(iii) $bab \sim_M ba$

(iv) $C_M(a) \cap [a]_{\sim_M} = \{a\}$

(v) $a \in Z(M) \rightarrow |[a]_{\sim_M}| = 1$

(vi) M is commutative if and only if every class with respect to $\sim_M$ consists of a single element only.

Proof: ad(i)-(iii): We calculate $ab(ba)ab = abbaab = abaab = abab = ab$ and $ba(ab)ba = baabba = babba = baba = ba$. Thus, $a(ba) \sim_M baa = ba$ and $(ba)b \sim_M bba = ba$ are valid, and we conclude $ab \sim_M ba \sim_M aba \sim_M bab$.

ad(iv)-(vi): Let every class with respect to $\sim_M$ be of order 1. For all $x, y \in M$ the identity $xy \sim_M yx$ is valid, and thus $xy = yx$ is true and M is commutative.

$\wedge$	e_1	e_2	e_3
e_1	e_1	e_2	e_3
e_2	e_2	e_2	e_2
e_3	e_3	e_3	e_3.

Table 1.1: structure constants of Π_2

Now let $ab = ba$ be valid and $a \sim_M b$. We derive $b = bab = abb = ab = aab = aba = a$, and thus $C_M(a) \cap [a]_{\sim_M} = \{a\}$ is true. Henceforth, for every $a \in Z(M)$ the class containing a is of order 1.$\diamond$

Example 1 For a field K we want to describe the nilradical, its factor algebra and its centralizer within $K\Pi_1$, $K\Pi_2$ and $K\Pi_3$.

(i) The algebra $K\Pi_1$ is – because of $\Pi_1 = \{(1)\}$ and corollary 4 – isomorphic to K. Its nilradical is the zero-subspace and its unique radical complement is $K\Pi_1$ which is self-centralizing.

(ii) The ordered set partitions of $\underline{2}$ are $e_1 := (12)$ (1-element within Π_2), $e_2 := (2, 1)$ and $e_3 := (1, 2)$. We present the structure constants of Π_2 (see table 1.1). We calculate $e_2 e_3 e_2 = e_2 e_2 = e_2$ and $e_3 e_2 e_3 e_3 e_3 = e_3$, and thus $e_2 \sim_{\Pi_2} e_3$ is true. By using part (i) of lemma 2 and corollary 5 we deduce $rad(K\Pi_2) = \langle e_2 - e_3 \rangle_K$. Let us define $T := \langle 1, e_2 \rangle_K$. T is a commutative subalgebra of $K\Pi_2$ with trivial intersection with the nilradical. Therefor, T is a radical complement. By using the commutativity of T we deduce $T \subseteq C_{K\Pi_2}(T) \subseteq K\Pi_2$. e_2 is not central in $K\Pi_2$, and henceforth (by a dimension argument) T is self-centralizing.

(iii) The 13 ordered set partitions of $\underline{3}$ are
$e_1 := (123)$ (1-element in Π_3),
$e_2 := (3, 12), e_3 := (2, 13), e_4 := (1, 23),$
$e_5 := (23, 1), e_6 := (13, 2), e_7 := (12, 3),$
$e_8 := (3, 2, 1), e_9 := (3, 1, 2), e_{10} := (2, 3, 1),$
$e_{11} := (2, 1, 3), e_{12} := (1, 3, 2), e_{13} := (1, 2, 3).$
We present the structure constants of $K\Pi_4$ within table 1.2 (the 1-element e_1 is omitted). The equivalence classes of Π_3 with respect to $\sim_{\Pi_3}$ can be determined by using the remark on page 9 in [18]: equivalent elements to $P := (P_1, \cdots, P_l)$ are completely derived from P b rearranging the components of P. In our context we get:

$\wedge$	e_2	e_3	e_4	e_5	e_6	e_7	e_8	e_9	e_{10}	e_{11}	e_{12}	e_{13}
e_2	e_2	e_8	e_9	e_8	e_9	e_2	e_8	e_9	e_8	e_8	e_9	e_9
e_3	e_{10}	e_3	e_{11}	e_{10}	e_3	e_{11}	e_{10}	e_{10}	e_{10}	e_{11}	e_{11}	e_{11}
e_4	e_{12}	e_{13}	e_4	e_4	e_{12}	e_{13}	e_{12}	e_{12}	e_{13}	e_{13}	e_{12}	e_{13}
e_5	e_8	e_{10}	e_5	e_5	e_8	e_{10}	e_8	e_8	e_{10}	e_{10}	e_8	e_{10}
e_6	e_9	e_6	e_{12}	e_9	e_6	e_{12}	e_9	e_9	e_9	e_{12}	e_{12}	e_{12}
e_7	e_7	e_{11}	e_{13}	e_{11}	e_{13}	e_7	e_{11}	e_{13}	e_{11}	e_{11}	e_{13}	e_{13}
e_8	e_8	e_8	e_8	e_8	e_8	e_8	e_8	e_8	e_8	e_8	e_8	e_8
e_9	e_9	e_9	e_9	e_9	e_9	e_9	e_9	e_9	e_9	e_9	e_9	e_9
e_{10}	e_{10}	e_{10}	e_{10}	e_{10}	e_{10}	e_{10}	e_{10}	e_{10}	e_{10}	e_{10}	e_{10}	e_{10}
e_{11}	e_{11}	e_{11}	e_{11}	e_{11}	e_{11}	e_{11}	e_{11}	e_{11}	e_{11}	e_{11}	e_{11}	e_{11}
e_{12}	e_{12}	e_{12}	e_{12}	e_{12}	e_{12}	e_{12}	e_{12}	e_{12}	e_{12}	e_{12}	e_{12}	e_{12}
e_{13}	e_{13}	e_{13}	e_{13}	e_{13}	e_{13}	e_{13}	e_{13}	e_{13}	e_{13}	e_{13}	e_{13}	e_{13}.

Table 1.2: structure constants of Π_3

$$[e_1]_{\sim_{\Pi_3}} = \{e_1\}$$
$$[e_2]_{\sim_{\Pi_3}} = \{e_2, e_7\},$$
$$[e_3]_{\sim_{\Pi_3}} = \{e_3, e_6\},$$
$$[e_4]_{\sim_{\Pi_3}} = \{e_4, e_5\} \text{ and}$$
$$[e_8]_{\sim_{\Pi_3}} = \{e_8, e_9, e_{10}, e_{11}, e_{12}, e_{13}\}.$$

Part (i) of lemma 2 and corollary 5 let us deduce that the 8-dimensional nilradical is exactly $\langle e_2 - e_7, e_3 - e_6, e_4 - e_5, e_8 - e_9, e_8 - e_{10}, e_8 - e_{11}, e_8 - e_{12}, e_8 - e_{13} \rangle_K$.

Every radical complement is – based on part (vi) of theorem 2 – isomorphic to K^5. We want to construct one radical complement. The subalgebra $\langle 1, e_2 \rangle_K$ is isomorphic to K^2. The centralizer of e_2 in $K\Pi_3$ is 5-dimensional. One element commuting with e_2 is the idempotent $e_3 - e_{10}$. In addition, $e_4 - e_{12}$ is an idempotent commuting with 1, e_2 and $e_3 - e_{10}$. Finally, e_8 is an idempotent within the centralizer of the set $\{1, e_2, e_4 - e_{12}, e_3 - e_{10}\}$. The centralizer of e_2 is 5-dimensional, and thus it is exactly the set $\langle 1, e_2, e_4 - e_{12}, e_3 - e_{10}, e_8 \rangle_K$. Following our construction of this centralizer we can deduce that it is commutative. In addition, it is K-linear generated by idempotent elements. Such elements are diagonalizable, and we can use theorem 5.5.1 in [22] to obtain the isomorphism to K^5. Based on our construction we also get that T is self-centralizing.$\diamond$

1.2 Open-ended questions

- Is remark 2 also valid for arbitrary finite idempotent monoids and thus the equivalence of corollaries 4 and 6 true, too?

- In view of theorems 1 and 2, determine the series of derivations of the associative algebra KM and its associated Lie-algebra KM° as well as the descending central-series of KM°! Is there a correspondent result for the decomposition of each member of these chains into local components? Are there connections between these chains and the classes of nilpotency and solvability?

- Is it true within theorem 2 that $(KM \circ KM) \cap T = KT \circ KT$ and similar intersection properties are valid?

1.3 Exercises

Exercise 2 *Let G be a finite p-group and K be a field of characteristic p. For all $g \in G$ calculate the minimal polynomial of g and $g - 1$ in KG.*

Exercise 3 *Let K be a field, A a K-algebra and e an idempotent of A. Calculate the minimal polynomial of e in A.*

Exercise 4 *Let K be a field, $n \in \mathbb{N}$ and $a, b \in \Pi_n$ mit $a \sim b$. Determine the minimal polynomial of $a - b$ and express $a - b$ as a sum of Lie-products with respect to the composition $\circ$.*

Exercise 5 *Let K be a field. Is Π_3 commutative? Is $K\Pi_4$ semisimple? Does $K\Pi_5$ split over K? Is $K\Pi_6$ separable?*

Exercise 6 *Visualize the $<$-lattice of Π_2 and Π_3 by using suitable graphs!*

Exercise 7 *How many elements within Π_3 are $<$ as, $>$ as and $\sim$ to $(12, 3)$?*

Exercise 8 *Let K be a field. Calculate the class of nilpotency of $rad(K\Pi_2)$ and $rad(K\Pi_3)$! Can you deduce a conjecture for this class for an arbitrary n?*

Exercise 9 *Let K be a field. Which dimension has $K\Pi_4$, its nilradical and the factor algebra by the nilradical? Prove that radical complements exist and determine their common dimension?*

Exercise 10 *Let* $P := (123, 456)$, $Q := (12, 3, 45, 6)$, $R := (123456)$ *and* $S := (1, 2, 3, 4, 5, 6)$. *Analyze the connections of* P, Q, R, S *with respect to* $<$, $>$ *and* $\sim$ *in* Π_6!

Exercise 11 *Let* K *be a field and* $n \in \mathbb{N}$. *Prove that* $K\Pi_n$ *and* D_n *have radical complements. Every radical complement of* D_n *(as a subalgebra of* $K\Pi_n$*) can be extended to a radical complement of* $K\Pi_n$. *(Tip: Use the theorem of Wedderburn-Malcev!)*

Exercise 12 *Let* K *be a field and* $n \in \mathbb{N}$. *Does* $K\Pi_4$ *possess a self-centralizing radical complement?*

Exercise 13 *Let* $P := (123, 456)$, $Q := (12, 3, 45, 6)$, $R := (123456)$ *and* $S := (1, 2, 3, 4, 5, 6)$. *Calculate* PQR, PR, $PQRS$, RPQ *and* SQS.

Exercise 14 *Let* K *be a field,* V *a finite-dimensional* K*-space and* B, C *two bases of* V. *Does a connection between* $Aug_B(V)$ *and* $Aug_C(V)$ *exist? Are these spaces identical, isomorphic, of the same dimension? What is the answer for a (semi)group algebra?*

Chapter 2

Idempotents, basis transformation and radical complements

This chapter contains the following basic facts about $K\Pi_n$:

- construction of a second basis of $K\Pi_n$ which is one of the key ideas in Manfred Schocker's article [18]

- description of the nilradical and of one special radical complement of $K\Pi_n$ by using the second basis (see also [18])

- description of all idempotents of $K\Pi_n$

- determination of the K-span of all idempotent of $K\Pi_n$

- evaluation of the K-span of all separable elements for the cases being one radical complement or the whole algebra.

2.1 Idempotents, basis transformation and radical complements

Definitions 3 We present some definitions and remarks which are already included in the article of Manfred Schocker ([18]). By Fin we symbolize the set of all finite subsets of $\mathbb{N}$. If $A \in Fin$, then the set Π_A of all ordered set partitions of A is a finite idempotent monoid with 1-element (A) with respect to the composition defined for $A = \underline{n}$ within the introduction. We symbolized this product by $\wedge_A$. For every $Q := (Q_1, \cdots, Q_k)$ the length of

Q is the number of components of Q: $l(Q) := k$.

Let $\Pi := \bigcup\limits_{A \in Fin} \Pi_A$. On Π we define the concatenation $\vee$ by

$$P \vee Q := \begin{cases} (P_1, \cdots, P_l, Q_1, \cdots, Q_k) & : & \bigcup\limits_{i=1}^{l} P_i \neq \bigcup\limits_{i=1}^{k} Q_i \\ 0 & : & otherwise \end{cases}$$

for all $P = (P_1, \cdots, P_l), Q = (Q_1, \cdots, Q_k) \in \Pi$. Let K be a field. By a linear extension argument we extend $\vee$ to a product on $K\Pi$.

Let $A \in Fin$. We define $\Pi_A^{\star} := \{(P_1, \cdots, P_l) \mid min(A) \in P_1\}$ and $e_A := \sum\limits_{P \in \Pi_A^{\star}} (-1)^{l(P)-1} P$. For all $Q = (Q_1, \cdots, Q_K) \in \Pi$ we set $e_Q := e_{Q_1} \vee \cdots \vee e_{Q_k}$.

Finally, we define $\Pi_A^{<} := \{(Q_1, \cdots, Q_k) \in \Pi_A \mid min(Q_1) < \cdots < min(Q_k)\}$.⋄

Manfred Schocker proves within [18] (see proposition 5.1, corollary 6.3 and theorem 5.4):

Theorem 3 *(Manfred Schocker) Let K be a field and $A \in Fin$.*

(i) $\{e_Q \mid Q \in \Pi_A\}$ *is a K-basis of $K\Pi_A$.*

(ii) $rad(K\Pi_n) = \langle e_Q - e_P \mid Q \in \Pi_n^{<}, P \in \Pi_n, P \sim_{\Pi_n} Q \rangle_K$

(iii) *The idempotents $e_T, T \in \Pi_A^{<}$ of $K\Pi_A$ are primitive and pairwise orthogonal and their K-span is a radical complement of $K\Pi_A$. In particular, every radical complement is commutative, splits over K and is of dimension $\mid \Pi_A^{<} \mid$.*

(iv) *The set $\Pi_A^{<}$ is a complete set of representatives for the $\sim$-classes of Π_A.*⋄

The next proposition contains the determination of all idempotents:

Proposition 2 *Let K be a field, A a finite-dimensional associative unitary K-algebra and $e_1, \cdots, e_n$ pairwise orthogonal idempotents of A such that $\langle e_1, \cdots, e_n \rangle_K$ is a radical complement of A.*

(i) *If e is an idempotent of $\langle e_1, \cdots, e_n \rangle_K$, then a subset $T \subseteq \underline{n}$ exists such that $e = \sum\limits_{t \in T} e_t$ is valid.*

(ii) *If e is an idempotent of A, then an element $r \in rad(A)$ and a subset $T \subseteq \underline{n}$ exist such that $e = (\sum\limits_{t \in T} e_t)^{1+r}$ is true..*

Proof: ad(i): The orthogonal vectors $e_1, \cdots, e_n$ are linear independent. Let e be an idempotent of $\langle e_1, \cdots, e_n \rangle_K$ and let $k_1, \cdots, k_n \in K$ such that $e = \sum\limits_{i=1}^{n} k_i e_i$ is valid. Based on the orthogonality and idempotency of $e_1, \cdots, e_n$ we calculate $\sum\limits_{i=1}^{n} k_i e_i = e = e^2 = \sum\limits_{i=1}^{n} k_i^2 e_i$. Thus, $k_i^2 = k_i$ and therefor $k_i \in \{0, 1\}$ is true for all $i \in \underline{n}$. We have proven (i).

ad(ii): Let e be an idempotent of A. The algebras $\langle e \rangle_K$ and K are isomorphic, and thus both are separable. Based on corollary 2.2.7 in [22] (enhanced version of the conjugacy part of the theorem of Wedderburn-Malcev) we find an element $r \in rad(A)$ such that $\langle e \rangle_K^{1+r} \le \langle e_1, \cdots, e_n \rangle_K$ is valid. e and the conjugate e^{1+r} are idempotent, and by using part (i) a subset $T \subseteq \underline{n}$ exists such that $e^{1+r} = \sum\limits_{t \in T} e_t$ is true. $1 + rad(A)$ is a subgroup of $E(A)$, and therefore an element $s \in rad(A)$ exists such that $(1 + r)^{-1} = 1 + s$ is valid. We calculate $e = (\sum\limits_{t \in T} e_t)^{1+s}$ which proves part (ii).$\diamond$

Corollary 7 *Let K be a field and $A \in Fin$. $K\Pi_A$ is the K-space-sum of all its radical complements.*

Proof: Based on theorem 3 we deduce that $K\Pi_A$ possesses a K-basis containing only idempotents. Every idempotent is contained in a radical complement by using proposition 2.$\diamond$

Remark 4 Let A be an associative finite-dimensional K-algebra with separable factor algebra by its nilradical. Using a similar argumentation as done in corollary 7 and proposition 2 (replace idempotent by separable element) we can prove that A has a K-basis containing only separable elements only if and only if A is the K-space-sum of all its radical complements.
The other extreme situation for the sum of all radical complements is to be exactly one radical complement. But by using the theorem of Wedderburn-Malcev the sum would be equal to every radical complement because they are conjugated and the sum is invariant under conjugation. Hence, this case is equivalent to the fact that exactly one radical complement exists. Within 2.4.5 in [22] this situation is analyzed: the nilradical and its unique radical complement centralize each other. If, in addition, A is solvable, then based on the same results the radical complement is central. In [21] this fact is characterized further: the associated Lie-algebra A° of A is nilpotent.$\diamond$

2.2 Open-ended questions

- Let K be a field and $n \in \mathbb{N}$. Determine the transformation matrix for the K-basis Π_n and $\{e_P \mid P \in \Pi_n\}$ of $K\Pi_n$!

- Let K be a field and $n \in \mathbb{N}$. For every $P \in \Pi_n$ an element $r \in rad(K\Pi_n)$ exists such that P^{1+r} resp. e_P^{1+r} is contained in a radical complement $\{e_P \mid P \in \Pi_n^<\}$ (theorem of Wedderburn-Malcev). Is it possible to construct such an element r?

- Let K be a finite field and $n \in \mathbb{N}$. How many idempotents are contained in $K\Pi_n$?

- Let K be a finite field and $n \in \mathbb{N}$. How many radical complements are contained in $K\Pi_n$?

- Let K be a field and M an idempotent monoid. Does a basis consisting of idempotent elements of KM exists such that a subset / submonoid of this basis is a basis of a radical complement?

2.3 Exercises

Exercise 15 *Let K be a finite field. How many and which idempotents are contained in $K\Pi_2$? Is the answer dependent on K?*

Exercise 16 *Prove remark 4!*

Exercise 17 *Let K be a finite field. How many and which idempotents are contained in $K\Pi_3$? Is the answer dependent on K?*

Exercise 18 *Let K be a finite field. How many and which radical complements are contained in $K\Pi_2$? Is the answer dependent on K?*

Exercise 19 *Let K be a finite field. How many and which radical complements are contained in $K\Pi_3$? Is the answer dependent on K?*

Exercise 20 *Determine the sets $\Pi_2^<$ and $\Pi_3^<$ and their order!*

Exercise 21 *Let K be a field. Determine the elements e_Q for all $Q \in \Pi_2$!*

Exercise 22 *Let K be a field. Determine the products $e_Q e_P$ for all $Q, P \in \Pi_2$!*

Exercise 23 *Let K be a field. Determine the elements e_Q for all $Q \in \Pi_3$!*

Exercise 24 *Let K be a field. Determine the products $e_Q e_P$ for all $Q, P \in \Pi_3$!*

Exercise 25 *Let K be a field. Calculate the transformation matrix for the basis Π_2 and $\{e_Q \mid Q \in \Pi_2\}$ of $K\Pi_2$!*

Exercise 26 *Let K be a field. Calculate the transformation matrix for the basis Π_3 and $\{e_Q \mid Q \in \Pi_2\}$ of $K\Pi_3$!*

Exercise 27 *Let K be a field. For all $Q \in \Pi_2$ construct an element $r \in rad(K\Pi_2)$ such that $Q^{1+r} \in \langle e_P \mid P \in \Pi_2^< \rangle_K$ is valid! Why does such an element r exist?*

Exercise 28 *Let K be a field. For all $Q \in \Pi_2$ find an element $r \in rad(K\Pi_2)$ such that $e_Q^{1+r} \in \langle e_P \mid P \in \Pi_2^< \rangle_K$! Why does such an element r exist?*

Exercise 29 *Let K be a field. For all $Q \in \Pi_3$ determine an element $r \in rad(K\Pi_3)$ such that $Q^{1+r} \in \langle e_P \mid P \in \Pi_3^< \rangle_K$ is true! Why does such an element r exist?*

Exercise 30 *Let K be a field. For all $Q \in \Pi_2$ determine an element $r \in rad(K\Pi_3)$ such that $e_Q^{1+r} \in \langle e_P \mid P \in \Pi_3^< \rangle_K$ is valid. Why does such an element r exist?*

Exercise 31 *Let $A \in Fin$ and K be a field. True or false:*

- *Π_A and $\Pi_{|A|}$ are isomorphic.*

- *$K\Pi_A$ and $K\Pi_{|A|}$ are isomorphic.*

Chapter 3

Dimensions

Within this chapter we focus on the following dimension-topics of $K\Pi_n$:

- dimension of $K\Pi_n$: counting ordered set partitions

- dimension of $(K\Pi_n)/(rad(K\Pi_n))$: counting unordered set partitions (Bell and Stirling numbers)

- dimension of $rad(K\Pi_n)$

- bounding these dimensions by using Solomons algebra

- growth of these dimensions.

3.1 Dimensions and lower bounds of the Solomon-Tits algebra and its radical factor algebra

Definitions 4 The free monoid over $\mathbb{N}$ is symbolized by $\mathbb{N}^\star$, and its elements are called words. If $q = q_1 \cdots q_k$ is such a word, then its length is noted by $\mid q \mid (:= k)$. For every number = letter $n \in \mathbb{N}$ we define the multi-degree of n with respect to q by $\mu_n(q) :=\mid \{i \mid 1 \leq i \leq\mid q \mid, n = q_i\}$.

Let $n \in \mathbb{N}$ and $q = q_1 \cdots q_k \in \mathbb{N}^\star$. q is called a decomposition of n – symbolized by $q \models n$ –, if $\sum_{i=1}^{k} q_i = n$ is valid. A decomposition q is a partition of n – denoted by $q \vdash n$ – if $q_1 \geq \cdots \geq q_k$ is true.

Two words v, w are called associated – marked by $v \approx w$ – if all letters n have the same multi-degree with respect to v and w: $\mu_n(v) = \mu_n(w)$.

If $n \in \mathbb{N}$ and $Q = (Q_1, \cdots, Q_k) \in \Pi_n$, then the type of Q is defined by the decomposition $Typ(Q) :=\mid Q_1 \mid \cdots \mid Q_k \mid$ of n.

For every $n \in \mathbb{N}$ let A_n be the alternating group of degree n.$\diamond$

38

Remark 5 Let K be a field and $n \in \mathbb{N}$. As stated within corollary 2, for every $n \in \mathbb{N}$ is Solomon's algebra D_n contained in $K\Pi_n$ as an isomorphic copy. The dimension of D_n is known to be 2^{n-1} – the number of decompositions of n –, and thus a lower bound for the dimension of the Solomon-Tits algebra is 2^{n-1}.

Another approach for this lower bound is the following fact: Let $n \in \mathbb{N}$ and $q = q_1 \cdots q_k$ be a decomposition of n, then we define the ordered set partition $Q_q := (Q_1, \cdots, Q_k)$ by $Q_1 := \underline{q_{1}}$ and $Q_i := \underline{q_1 + \cdots + q_i} \setminus \underline{q_1 + \cdots + q_{i-1}}$ for all $2 \leq i \leq k$. The function $q \mapsto Q_q$ is injective.

A better lower bound for the dimension of $K\Pi_n$ is $n!$: For every permutation $\alpha \in S_n$ we define the ordered set partition $P_\alpha := (1\alpha, \cdots, n\alpha)$. Also this function $\alpha \mapsto P_\alpha$ is injective. The image of this function is a class with respect to $\sim_{\Pi_n}$.$\diamond$

A simple combinatorial argument shows us:

Remark 6 *Every word q possesses exactly* $\dfrac{|q|!}{\prod\limits_{a \in \mathbb{N}} \mu_q(a)!}$ *associated words.*$\diamond$

Theorem 4 *Let $n \in \mathbb{N}$.*

(i) $|\ \Pi_n\ | = n! \cdot \sum\limits_{q_1 \cdots q_k \models n} \dfrac{1}{\prod\limits_{i=1}^{k} (q_i)!}$

(ii) $|\ \Pi_n\ | = n! \cdot \sum\limits_{q_1 \cdots q_k \vdash n} \dfrac{|q|!}{\prod\limits_{a \in \mathbb{N}} \mu_q(a)! \prod\limits_{i=1}^{k} (q_i)!}$

<u>Proof:</u> ad(i): The symmetric group S_n acts by $(P_1, \cdots, P_k)\alpha := (P_1\alpha, \cdots, P_k\alpha)$ on Π_n $(P_i\alpha := \{x\alpha \mid x \in P_i\})$. Let $P = (P_1, \cdots, P_k) \in \Pi_n$. The orbit of P under S_n contains exactly those ordered set partitions of type P. The stabilizer of P under S_n is a Young-subgroup which is isomorphic to $S_{|P_1|} \times \cdots \times S_{|P_k|}$.[1] Thus, we have proven part (i).

ad(ii): Every association-class contains exactly one partition. The denominator of each summand of the formula in (i) is identical for every associated word. By using remark 6 and (i) we derive (ii).$\diamond$

---

[1] Alfred Young

$p \vdash 5$	$\mid [p]_\approx \mid$	$\prod\limits_{i=1}^{k} (p_i)!$
5	1	5!
41	2	4!
311	3	3!
32	2	3!2!
221	3	2!2!
2111	4	2!
11111	1	1.

Table 3.1: table for $\mid \Pi_5 \mid$

n	2^{n-1}	$n!$	$\mid \Pi_n \mid$
1	1	1	1
2	2	2	3
3	4	6	13
4	8	24	70
5	16	120	541.$\diamond$

Table 3.2: lower bounds for $\mid \Pi_n \mid$

Example 2 We prove that exactly 541 ordered set partitions of $\underline{5}$ exist. For this, we use theorem 4 and table 3.1 to obtain $\mid \Pi_5 \mid = 5!(1 + 4\frac{1}{2!} + 3\frac{1}{2!2!} + 2\frac{1}{3!2!} + 3\frac{1}{3!} + 2\frac{1}{4!} + 1\frac{1}{5!}) = 120 + 240 + 90 + 20 + 60 + 10 + 1 = 541$. Within table 3.2 we present lower bounds based on remark 5 for the dimension of $K\Pi_n$ for $n \in \underline{5}.\diamond$

Remark 7 Let K be a field and $n \in \mathbb{N}$. On page 10 in [18] it is proven that the dimension of the radical factor algebra of $K\Pi_n$ is exactly the number of unordered set partitions of $\underline{n}$. By using corollary 2 we deduce that Solomon's algebra D_n is isomorphic to a subalgebra T of $K\Pi_n$. $K\Pi_n$ is based on theorem 2 solvable, and thus we use proposition 5 in [21] to obtain $rad(T) \leq rad(K\Pi_n)$. Therefor, $D_n/rad(D_n)$ is isomorphic to a subalgebra of $(K\Pi_n)/(rad(K\Pi_n))$. In particular, $dim_K(K\Pi_n/rad(K\Pi_n)) \geq dim_K(D_n/rad(D_n))$ is valid. This lower bound is well-known to be $p(n)$: the number of partitions of $n.\diamond$

40

Definitions 5 Let $n \in \mathbb{N}_0$ and $0 \leq k \leq n$. The Bell-numbers $B(n)$[2] are defined as the quantity of unordered set partitions of $\underline{n}$. Thus, $B(n)$ is the K-dimension of the radical factor algebra of $K\Pi_n$. The k-th[3] Stirling-number (of second kind) $S(n, k)$ is the number of unordered set partitions of $\underline{n}$ consisting of exactly k subsets of $\underline{n}$.$\diamond$

We present some well-known results for the Bell- and Stirling-numbers (see [31]):

Theorem 5 *Let* $n \in \mathbb{N}_0$ *and* $0 \leq k \leq n$.

(i) $S(n, k) = S(n - 1, k - 1) + k \cdot S(n - 1, k)$

(ii) $S(n, k) = \sum\limits_{r=0}^{k} (-1)^{k-r} \frac{r^n}{r!(k-r)!}$

(iii) $B(n) = \sum\limits_{i=1}^{n} S(n, k)$

(iv) $B(n + 1) = \sum\limits_{k=0}^{n} \binom{n}{k} \cdot B(k)$

(v) $B(n) = \frac{1}{e} \sum\limits_{k=0}^{\infty} \frac{k^n}{k!}$ *(Dobinsky).*$\diamond$

Remark 8 Let $n \in \mathbb{N}$. Bell-numbers can be calculated recursively by using the following scheme (see [31]):

(1.) Start with the number 1. This is also the first column.
(2.) Start a new column on the left side with the rightmost number of the previous column.
(3.) Calculate the next number in the column as the sum of the number to the left and to the top.
(4.) Repeat step 3 as long as the quantity of numbers in the column is exactly the quantity of numbers within the previous column plus one.

---

[2]Eric Temple Bell

[3]James Stirling

(1.)	**1**	–	–	–	–
(2.)	1	**2**	–	–	–
(3.)	2	3	**5**	–	–
(4.)	5	7	10	**15**	–
(5.)	15	20	27	37	**52.**

Table 3.3: triangle-scheme for counting Bell-numbers

n	$p(n)$	$B(n)$
1	1	1
2	2	2
3	3	5
4	5	15
5	7	52.◇

Table 3.4: partitions and Bell-number

(5.) The rightmost number in column i is the i-th Bell-number.

We calculate the Bell-numbers $B(n)$ for $n = 1, \cdots, 5$ within table 3.3 and list the corresponding partition-numbers within table 3.4.◇

Corollary 8 *Let $n \in \mathbb{N}$ and K be a field.*

(i) $\forall k \in \underline{n}_0 : |\, \{P \in \Pi_n \mid l(P) = k\} \,| = k! \cdot S(n, k)$
In particular, there are exactly $n!$ ordered set partitions of length n.

(ii) $|\, \Pi_n \,| = \sum\limits_{k=0}^{n} k! \cdot S(n, k)$

(iii) $|\, \Pi_{n+1} \,| - |\, \Pi_n \,| = 2 \cdot \sum\limits_{k=0}^{n} k \cdot k! \cdot S(n, k)$

(iv) $|\, \Pi_{n+1} \,| \geq 3 \,|\, \Pi_n \,| \geq 3^n$

(v) $dim_K(rad(K\Pi_n)) = \sum\limits_{k=0}^{n} (k! - 1) \cdot S(n, k)$

(vi) $dim_K(\Pi_n / rad(K\Pi_n)) = B(n)$.

Proof: ad(i): Let $k \in \underline{n}_0$. We focus on the surjection $\varphi : (P_1, \cdots, P_k) \mapsto \{P_1, \cdots, P_k\}$ from the set of ordered on the set of unordered set partitions

42

of length k. Every unordered set partitions has exactly $k!$ pre-images under φ.

ad(ii): This statement is a direct consequence of part (i).

ad(iii): For all $k \in \underline{n}_0$ we use theorem 5 to obtain the recursion-formula $S(n, k) = S(n - 1, k - 1) + k \cdot S(n - 1, k)$. Using this and part (ii) we can calculate:

$$
\begin{aligned}
& \mid \Pi_{n+1} \mid - \mid \Pi_n \mid \\
= & \sum_{k=0}^{n+1} k! \cdot S(n + 1, k) - \sum_{k=0}^{n} k! \cdot S(n, k) \\
= & \sum_{k=0}^{n} k! \cdot S(n, k - 1) + (\sum_{k=0}^{n} k! \cdot k \cdot S(n, k)) + (n + 1)! - \sum_{k=0}^{n} k! \cdot S(n, k) \\
= & \sum_{k=0}^{n-1} (k + 1)! \cdot S(n, k) + (\sum_{k=0}^{n} k! \cdot k \cdot S(n, k)) + (n + 1)! - \sum_{k=0}^{n} k! \cdot S(n, k) \\
= & \sum_{k=0}^{n} (k + 1)! \cdot S(n, k) + \sum_{k=0}^{n} k! \cdot k \cdot S(n, k) - \sum_{k=0}^{n} k! \cdot S(n, k) \\
= & \ 2 \cdot \sum_{k=0}^{n} k \cdot k! \cdot S(n, k).
\end{aligned}
$$

ad(iv): This part is deductable from parts (ii) and (iii).

ad(v)+(vi): By using remark 7 the dimension of the radical factor algebra of $K\Pi_n$ is exactly $B(n)$. Thus, the dimension of the nilradical is given by $\mid \Pi_n \mid - B(n)$. The difference is calculated within theorem 5 and part (ii) as desired.$\diamond$

Remark 9 Let $n \in \mathbb{N}$ and K be a field. The Bell-numbers $B(n)$ are identified as K-dimension of the radical factor algebra of $K\Pi_n$. They also occur during the analysis of counting all unital subalgebras within one radical complement of $K\Pi_n$. A radical complement T (and also all based on theorem of Wedderburn-Malcev) is isomorphic to the algebra K^l such that $l := \mid \Pi_A^{<} \mid = B(n)$ is valid (see theorem 3). By using exercise 8, page 84 in [17] is the number of unital subalgebras of T exactly $B(l) = B(B(n))$.$\diamond$

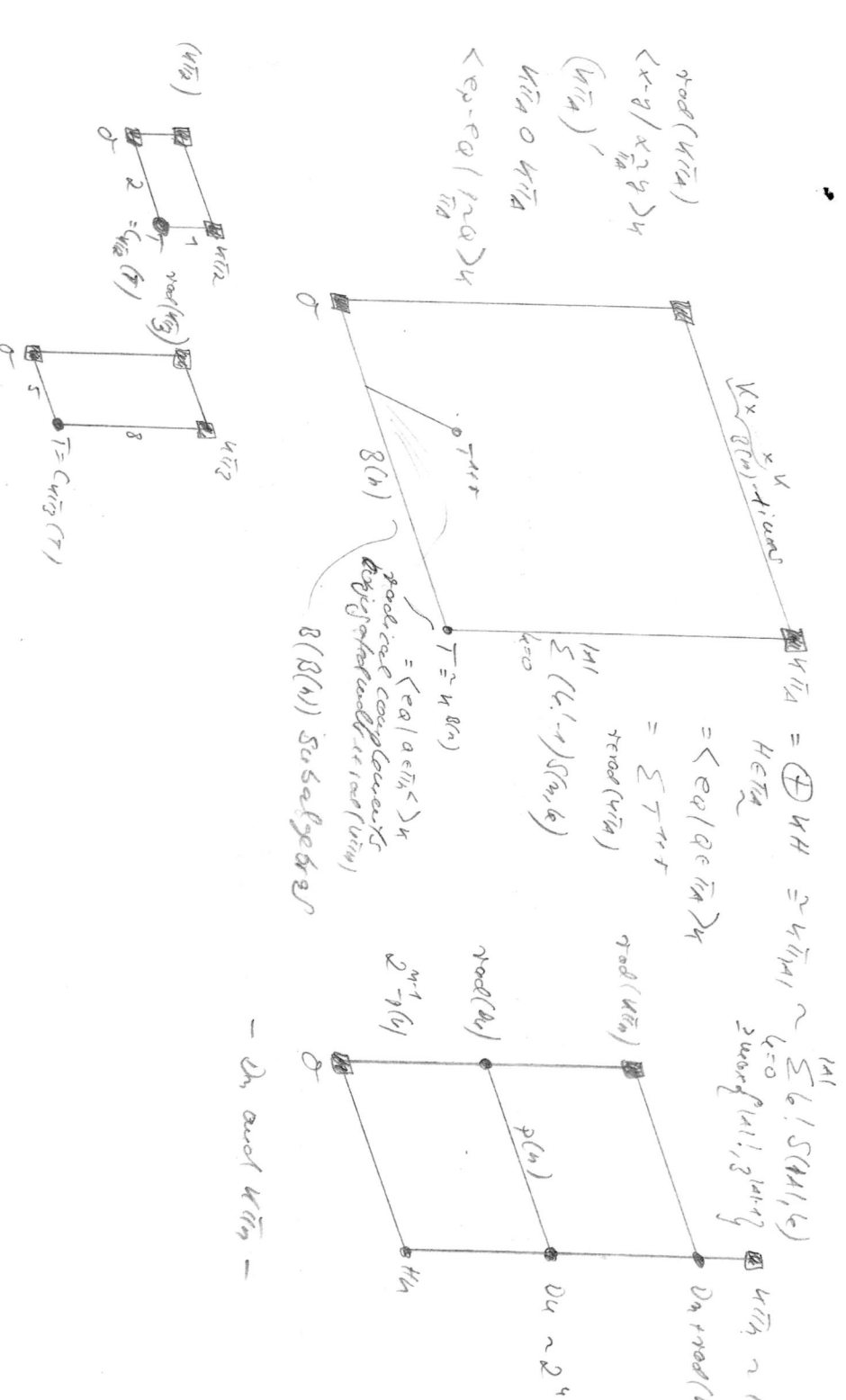

· Chapter 1-3 ·

$\mathcal{H}_a = \bigoplus_{H \in \overline{I}_a} \mathcal{H}_{[H]}$, $\mathcal{H}_a \supset \sum_{b=0}^{|H|} b! \, S(|H|, b)$

$\operatorname{grad}(\mathcal{H}_a)$

$\langle x \cdot y \mid x_2^2 y_2 \rangle_{\overline{I}_a} \geq 4$

$(\mathcal{H}_a)'$

$\mathcal{H}_a \otimes \mathcal{H}_a$

$\langle e_2 \cdot e_2 \mid e_2 \cdot e_2 \rangle_{\overline{I}_a} \geq 4$

$K \times (\overline{I}_a) \rightarrow \lambda_{twn}$

$\overline{K}_a \, (\overline{I}_a) \rightarrow \lambda_{twn}$

$= \langle e_2 \mid e_2 \in \overline{I}_a \rangle_K$

$= \sum_T \lambda_{twn} T$

$= \sum_{b=0}^{|H|} b! \, S(|H|, b)$

$\sum_{b=0}^{|H|} (b!-1) \, S(|H|, b)$

$\operatorname{grad}(\mathcal{H}_{twn})$

$T = \lambda \, B(n)$

$\int T = \lambda \, B(n)$

$B(h)$

$\langle e_2 \mid e_2 \in \overline{I}_a \rangle_K$

$= \langle e_2 \mid e_2 \in \overline{I}_a \rangle_K$

primitive components
trigonometric coefficients

$B(B(h))$ Subalgebras

$\operatorname{grad}(b_n)$ $2^{n-1}(n)$

$p(n)$

B_n

$\mathcal{H}_a \sim 2^n$

$B_n + \operatorname{grad}(\mathcal{H}_{twn})$

$\mathcal{H}_a \sim l \overline{I}_a$

— B_n and $\mathcal{H}_{twn}$ —

$(\mathcal{H}_{\overline{I}_2})$

$\operatorname{grad}(B_3)$

$= C_{n_{12}}(\pi)$

$\overline{\mathcal{H}}_{\overline{I}_2}$

$\overline{\mathcal{H}}_{\overline{I}_2}$

$T = C_{n_{12}}(\pi)$

— Diminutions of Prebial's Prebial components —

3.2 Open-ended questions

- Let $n \in \mathbb{N}$ and K be a field. Do the numbers $dim_K(K\Pi_{n+1}) - dim_K(K\Pi_n)$ converge or diverge?

- Let $n \in \mathbb{N}$ and K be a field. Do the numbers $\frac{dim_K(K\Pi_{n+1})}{dim_K(K\Pi_n)}$ converge or diverge?

- Let $n \in \mathbb{N}$ and K be a field. Do the numbers $dim_K(K\Pi_n) - dim_K(D_n)$ converge or diverge?

- Let $n \in \mathbb{N}$ and K be a field. Do the numbers $\frac{dim_K(K\Pi_n)}{dim_K(D_n)}$ converge or diverge?

- Let $r, n \in \mathbb{N}$. Does a formula exists for $B^r(n) = \underbrace{B(...(n))}\, r - times$?

- Let $n \in \mathbb{N}$ and K be a field. We focus on D_n as subalgebra of $K\Pi_n$. Does a basis of $K\Pi_n$ exist such that s subset of it is a basis for D_n? What is the answer with respect to radical complements of $K\Pi_n$ and D_n? Does a basis exist which can handle both topics simultaneously? Is it possible to extend the result to arbitrary monoid algebras or even associative algebras?

3.3 Exercises

Exercise 32 *Determine all partitions of* 4 *and count them.*

Exercise 33 *Determine all decompositions of* 5 *and count them.*

Exercise 34 *Is the word* $1.3.4.5.8^4$ *a decomposition or partitions of* 21*? What is the length of this word?*

Exercise 35 *Are the words* 1.3.4.5.8 *and* 4.3.8.5.1 *associated based on the alphabet* $\mathbb{N}$*? Which are the associated words of these words? Count these associated words! What is the type of* $(8.9.1, 2.3.4.5, 6.7)$*? Prove that this type is not associated to* 1.3.4.5.8*!*

Exercise 36 *To be associated is an equivalence relation on the set of words for an arbitrary alphabet.*

---

[4]The elements of $\mathbb{N}$ are separated by the symbol '.' within a word over $\mathbb{N}^\star$. E.g., 1.2.4 consists of the letters 1, 2 and 4 and the word 1.24 is determined by the letters 1 and 24. For a more transparent visualization also 1.(24) is used instead of 1.24.

Exercise 37 *For which and for how many associated elements of 1.3.4.5.8 are the following statements true or false:*

(i) The multi-degree of 3 is 2.

(ii) The multi-degree of 5 is 3.

(iii) The multi-degree of 1 is 0.

Exercise 38 *Prove remark 5!*

Exercise 39 *Prove remark 6!*

Exercise 40 *Prove theorem 5!*

Exercise 41 *Prove remark 9!*

Exercise 42 *Let K be a field. For $n = 6, 7, 8, 9, 10$ calculate 2^{n-1}, $n!$, $\mid \Pi_n \mid$, $\mid \Pi_n \mid - \mid \Pi_{n-1} \mid$ and $\frac{\mid \Pi_n \mid}{\mid \Pi_{n-1} \mid}$. In what way are these numbers connected to the structure of $K\Pi_n$?*

Exercise 43 *Let K be a field. For $n = 6, 7, 8, 9, 10$ determine $p(n)$, $B(n)$, $\mid \Pi_n \mid$ and $\mid \Pi_n \mid - B(n)$. In what way are these numbers connected to the structure of $K\Pi_n$?*

Chapter 4

Left and right ideals

Within this chapter we answer the following questions related to left and right ideals of $K\Pi_n$:

- Do special embeddings of $K\Pi_n$ in $K\Pi_{n+1}$ exist and what is their relation to right and left ideals?

- Is $K\Pi_n$ a (Quasi)-Frobenius algebra?

- Is $K\Pi_n$ uniserial?

- Is $K\Pi_n$ local?

- Is $K\Pi_n$ isomorphic to group algebra of a symmetric group?

- Does a special chain of ideals of $K\Pi_n$ exist which is based on the length-function?

- Do principal ideals in $K\Pi_n$ exists which are no principal right or left ideals (and the other combinations, too)?

- What is the connection of $K\Pi_n$ to Duo-algebras?

4.1 Embeddings

Remark 10 Let $n \in \mathbb{N}$. The injections (sending 1 to 1) from Π_n into Π_{n+1} defined by

$$\alpha : (P_1, \cdots, P_k) \mapsto (P_1 \cup \{n+1\}, P_2, \cdots, P_k) \text{ and}$$

$$\beta : (P_1, \cdots, P_k) \mapsto (P_1, \cdots, P_{k-1}, P_k \cup \{n+1\})$$

48

are – for $n \geq 2$ – no semigroup homomorphism because

$$
\begin{aligned}
& ((1, \cdots, n) \wedge (\underline{n-1}, n))\alpha \\
= {} & (1, \cdots, n)\alpha \\
= {} & (1(n+1), \cdots, n) \\
\neq {} & (n+1, 1, 2, \cdots, n) \\
= {} & (1(n+1), \cdots, n) \wedge (\underline{n-1} \cup \{n+1\}, n) \\
= {} & (1, \cdots, n)\alpha \wedge (\underline{n-1}, n)\alpha
\end{aligned}
$$

and

$$
\begin{aligned}
& ((1, \cdots, n) \wedge (n, \underline{n-1}, n))\beta \\
= {} & (1, \cdots, n)\beta \\
= {} & (1(n+1), \cdots, n) \\
\neq {} & (1, 2, \cdots, n, n+1) \\
= {} & (1, \cdots, n(n+1)) \wedge (n, \underline{n-1} \cup \{n+1\}) \\
= {} & (1, \cdots, n)\beta \wedge (n, \underline{n-1})\beta
\end{aligned}
$$

are valid.$\diamond$

Proposition 3 *Let $n \in \mathbb{N}$. The reflection*

$$
s_n : (P_1, P_2, \cdots, P_{k-1}, P_k) \mapsto (P_k, P_{k-1}, \cdots, P_2, P_1)
$$

is an involution of Π_n which is length-invariant.

<u>Proof:</u> We have only to prove that s_n is an homomorphism. Let $P := (P_1, \cdots, P_k)$ and $Q := (Q_1, \cdots, Q_l)$ two elements of Π_n. We calculate:

$$
\begin{aligned}
& (P \wedge Q)s_n \\
= {} & (P_1 \cap Q_1, \cdots, P_1 \cap Q_l, P_2 \cap Q_1, \cdots, P_2 \cap Q_l, \\
& \vdots \\
& P_{k-1} \cap Q_1, \cdots, P_{k-1} \cap Q_l, P_k \cap Q_1, \cdots, P_k \cap Q_l)s_n \\
= {} & (P_k \cap Q_l, \cdots, P_1 \cap Q_l, P_{k-1} \cap Q_l, \cdots, P_{k-1} \cap Q_1, \\
& \vdots \\
& P_2 \cap Q_l, \cdots, P_2 \cap Q_1, P_1 \cap Q_l, \cdots, P_1 \cap Q_1) \\
= {} & (P_k, P_{k-1}, \cdots, P_2, P_1) \wedge (Q_l, Q_{l-1}, \cdots, Q_2, Q_1) \\
= {} & Ps_n \wedge Qs_n. \diamond
\end{aligned}
$$

Proposition 4 *Let $n \in \mathbb{N}$. The functions*

$$r_n : (P_1, \cdots, P_k) \mapsto (P_1, \cdots, P_k, n+1) \text{ and}$$

$$l_n : (P_1, \cdots, P_k) \mapsto (n+1, P_1, \cdots, P_k)$$

are semigroup monomorphism from Π_n into Π_{n+1} such that $l_n = s_n r_n s_{n+1}$ is true.

In particular, $\Pi_n r_n$ and $\Pi_n l_n$ are isomorphic to Π_n and are sub-semigroups of Π_{n+1}.

Proof: We have only to prove that r_n and l_n are homomorphism. Let $P := (P_1, \cdots, P_k)$ and $Q := (Q_1, \cdots, Q_l)$ be two elements of Π_n. We calculate:

$$
\begin{aligned}
&(P \wedge Q)r_n \\
= \ &(P_1 \cap Q_1, \cdots, P_1 \cap Q_l, P_2 \cap Q_1, \cdots, P_2 \cap Q_l, \\
&\vdots \\
&P_{k-1} \cap Q_1, \cdots, P_{k-1} \cap Q_l, P_k \cap Q_1, \cdots, P_k \cap Q_l)r_n \\
= \ &(P_1 \cap Q_1, \cdots, P_1 \cap Q_l, P_2 \cap Q_1, \cdots, P_2 \cap Q_l, \\
&\vdots \\
&P_{k-1} \cap Q_1, \cdots, P_{k-1} \cap Q_l, P_k \cap Q_1, \cdots, P_k \cap Q_l, n+1)
\end{aligned}
$$

and

$$
\begin{aligned}
&Pr_n \wedge Qr_n \\
= \ &(P_1, \cdots, P_k, n+1) \wedge (Q_1, \cdots, Q_l, n+1) \\
= \ &(P_1 \cap Q_1, \cdots, P_1 \cap Q_l, \emptyset, P_2 \cap Q_1, \cdots, P_2 \cap Q_l, \emptyset, \\
&\vdots \\
&P_{k-1} \cap Q_1, \cdots, P_{k-1} \cap Q_l, \emptyset, P_k \cap Q_1, \cdots, P_k \cap Q_l, \emptyset, \emptyset, \cdots, \emptyset, n+1).
\end{aligned}
$$

Empty sets can be omitted by definition and thus r_n is a homomorphism. By using $l_n = s_n r_n s_{n+1}$ and proposition 3 we finish the proof.$\diamond$

Theorem 6 *Let K be a field and $n \in \mathbb{N}$.*

(i) $K\Pi_n r_n = (\underline{n}, n+1)K\Pi_{n+1} \cong K\Pi_n$

(ii) $K\Pi_n l_n = (n+1, \underline{n})K\Pi_{n+1} \cong K\Pi_n$

(iii) $K\Pi_{n+1} = K\Pi_n r_n \oplus (1 - (\underline{n}, n+1))K\Pi_{n+1}$

(iv) $K\Pi_{n+1} = K\Pi_n \, l_n \oplus (1 - (n+1, \underline{n}))K\Pi_{n+1}$

(v) $\{(P_1, \cdots P_k, n+1) \mid k \geq 1, (P_1, \cdots, P_k) \in \Pi_n\}$ *is a K-basis of the right ideal $K\Pi_n \, r_n$.*

(vi) $\{(n+1, P_1, \cdots P_k) \mid k \geq 1, (P_1, \cdots, P_k) \in \Pi_n\}$ *is a K-basis of the right ideal $K\Pi_n \, l_n$.*

(vii) $\{P - (\underline{n}, n+1)P \mid P = (P_1, \cdots, P_k) \in \Pi_{n+1}, P_k \neq \{n+1\}\}$ *is a K-basis of $(1 - (\underline{n}, n+1))K\Pi_{n+1}$ consisting of $2 \sum\limits_{k=0}^{n} k \, k! \, S(n,k)$ elements.*

(viii) $\{P - (n+1, \underline{n})P \mid P = (P_1, \cdots, P_k) \in \Pi_{n+1}, P_1 \neq \{n+1\}\}$ *is a K-basis of $(1 - (n+1, \underline{n}))K\Pi_{n+1}$ consisting of $2 \sum\limits_{k=0}^{n} k \, k! \, S(n,k)$ elements.*

(ix) $K\Pi_n \, r_n \cap K\Pi_n \, l_n = \{0\}$
In particular, $K\Pi_n \, l_n \oplus K\Pi_n \, r_n$ is a right ideal of K-dimension $2 \cdot |\Pi_n|$.

Proof: Let $(P_1, \cdots, P_r, \cdots, P_l) \in \Pi_{n+1}$ such that $n+1 \in P_r$ is valid. We calculate
$(\underline{n}, n+1) \wedge (P_1, \cdots, P_r, \cdots, P_l) = (P_1, \cdots, P_r \setminus \{n+1\}, \cdots, P_l, n+1)$ and
$(n+1, \underline{n}) \wedge (P_1, \cdots, P_r, \cdots, P_l) = (n+1, P_1, \cdots, P_r \setminus \{n+1\}, \cdots, P_l)$.
Thus, $(\underline{n}, n+1)K\Pi_{n+1} \subseteq K\Pi_n \, r_n = (\underline{n}, n+1)K\Pi_n \, r_n \subseteq (\underline{n}, n+1)K\Pi_{n+1}$ and $(n+1, \underline{n})K\Pi_{n+1} \subseteq K\Pi_n \, l_n = (n+1, \underline{n})K\Pi_n \, l_n \subseteq (n+1, \underline{n})K\Pi_{n+1}$ are true. Now we use proposition 4 to deduce parts (i) and (ii).
The statements (iii) and (iv) are a direct consequence of parts (i) and (ii) and parts (vii) and (viii) can be deducted by using parts (iii) and (iv) as well as corollary 8.
Proposition 4 is used to obtain (v) and (vi).
Statement (ix) is a consequence of parts (v) and (vi): the basis are disjoint because there is no ordered set partition of length ≥ 2 such that its first and last component is identical to $\{n+1\}$.$\diamond$

4.2 Principal ideals

Proposition 5 *Let K be a field and $n \in \mathbb{N}_{\geq 2}$. The following statements are valid:*

(i) $K\Pi_n \, e_{(\underline{n})}$ *is a left but no right ideal.*

(ii) $K\Pi_n \, r_n$ *is a right but no left ideal.*

Proof: Following page 22 in [18] we obtain that $K\Pi_n\,e_{(\underline{n})}$ is a left ideal of dimension $l((\underline{n}))! = 1$. If $K\Pi_n\,e_{(\underline{n})}$ would be a right ideal, then it would contain the right ideal $e_{(\underline{n})}\,K\Pi_n$. But this is – based on remark 6.5 in [18] – of dimension $2\cdot\mid\Pi_{n-1}\mid\,\geq 2$. Thus, $K\Pi_n\,e_{(\underline{n})}$ is no right ideal.

Now we prove that $K\Pi_n\,r_n = (\underline{n}, n+1)K\Pi_{n+1}$ (see theorem 6) is no left ideal. $(n+1,\underline{n})\wedge(\underline{n}, n+1) = (n+1,\underline{n})$ is valid, and $(n+1,\underline{n})$ is – by using part (v) of theorem 6 – not contained in $K\Pi_n\,r_n$.$\diamond$

Remark 11 We remark the following topics in view of proposition 5: The right ideal $K\Pi_n\,r_n$ is contained proper in the left ideal $K\Pi_{n+1}(n+1,\underline{n})$. This left ideal is an ideal and is spanned by the ordered set partitions $(P_1,\cdots,P_l)$ such that $l\geq 2$ and for at least one $2\leq i\leq l$ the identity $P_i = \{n+1\}$ is true. The ideal $K\Pi_{n+1}(n+1,\underline{n})$ is direct to the right ideal $K\Pi_n\,l_n$. The sum of these two substructures is again an ideal: it is spanned by the ordered set partitions $(P_1,\cdots,P_l)$ such that at least one P_i is equal to $\{n+1\}$ for one index i and its length is at least 2.$\diamond$

Corollary 9 *Let K be a field and $n\in\mathbb{N}$. The following statements are equivalent:*

(i) $n = 1$

(ii) Π_n possesses a single element only.

(iii) The associative algebras $K\Pi_n$ and K are isomorphic.

(iv) $K\Pi_n$ is a division algebra.

(v) $K\Pi_n$ is simple.

(vi) $K\Pi_n$ is local.

(vii) $K\Pi_n$ is separable.

(viii) $K\Pi_n$ is semisimple.

(ix) $K\Pi_n$ is commutative.

(x) Every class of Π_n with respect to $\sim_{\Pi_n}$ is of order 1.

Proof: see corollaries 3, 4, 6 and proposition 5.$\diamond$

Proposition 6 *Let $n\in\mathbb{N}$ and K be a field.*

(i) For all $P,Q\in\Pi_n$ the identity $\max\{n, l(P)l(Q)\}\geq l(P\wedge Q)\geq \max\{l(P), l(Q)\}$ is valid.

(ii) $(1,23) \wedge (12,3) = (1,2,3)$, $(14,23) \wedge (12,34) = (1,4,2,3)$, $(1,2) \wedge (1,2) = (1,2)$ *(These examples illustrate (i).)*

(iii) *For all* $k \in \underline{n+1}$, *we define* $I_{\geq k} := \langle \{P \in \Pi_n \mid l(P) \geq k\} \rangle_K$. *The chain* $(0 = I_{\geq n+1}, I_{\geq n}, \cdots, I_{\geq 2}, I_{\geq 1} = K\Pi_n)$ *consists of ideals of* $K\Pi_n$.

(iv) *For all* $k \in \underline{n}$, *the relation* $dim_K(I_{\geq k}/I_{\geq k+1}) = k! S(n,k)$ *is valid.*

Proof: ad(i): Let $P := (P_1, \cdots, P_k)$ and $Q := (Q_1, \cdots, Q_l)$ two elements of Π_n. We calculate:

$$
\begin{aligned}
&P \wedge Q \\
= \; &(P_1 \cap Q_1, \cdots, P_1 \cap Q_l, P_2 \cap Q_1, \cdots, P_2 \cap Q_l, \\
&\vdots \\
&P_{k-1} \cap Q_1, \cdots, P_{k-1} \cap Q_l, P_k \cap Q_1, \cdots, P_k \cap Q_l).
\end{aligned}
$$

For every $i \in \underline{n}$ the identity $P_i \cap (Q_1 \cup \cdots \cup Q_l) = P_i$ is valid. Thus, at least one index $j \in \underline{n}$ exists such that $P_i \cap Q_j \neq \emptyset$ is valid. (In every column of the product of P and Q at least one entry different from $\emptyset$ exists.) Likewise, for every $j \in \underline{n}$ at least one $i \in \underline{n}$ exists such that $P_i \cap Q_j \neq \emptyset$ is true. (In every column of the product of P and Q at least one entry different from $\emptyset$ exists.) Thus, part (i) is proven.

ad(ii): This statement can be verified by a straightforward calculation.

ad(iii): This part is a direct consequence of part (i).

ad(iv): Let $k \in \underline{n}$. The K-space $I_{\geq k}/I_{\geq k+1}$ is isomorphic to the K-space which is spanned by all ordered set partitions of length k. Corollary 8 already determines the desired dimension.$\diamond$

Theorem 7 *Let* $n \in \mathbb{N}$ *and* K *be a field.*

(i) $dim_K(I_{\geq n}) = n!$

(ii) $I_{\geq n} = K\Pi_n e_{(1,2,\cdots,n-1,n)} = K\Pi_n (1,2,\cdots,n-1,n)$
$(e_{(1,2,\cdots,n-1,n)} = (1,2,\cdots,n-1,n))$

(iii) *Let* $P, Q \in \Pi_n$ *and* $l(P) = n$. *Then,* $P \wedge Q = P$ *is valid. In particular,* $I_{\geq n}$ *is a principal ideal and only for* $n = 1$ *a principal right ideal. Thus,* $K\Pi_n$ *is only for* $n = 1$ *uniserial.*

(iv) $K\Pi_n$ *does not possess for* $r \geq 4$ *a subalgebra isomorphic to* KS_r.

(v) *The group of units of* $K\Pi_3$ *possesses for* $char(K) = 3$ *a subgroup isomorphic to* S_3 *which consists of linear dependent vectors.*

(vi) *Only for* $n = 1$ *and* $n = 2, char(K) = 2$ *the algebras* $I_{\geq n}$ *and* KS_n *are isomorphic. For* $n = 2, char(K) \neq 2$ *the radical complements of* $K\Pi_2$ *are isomorphic to* KS_2.

(vii) $I_{\geq n}$ *is a completely reducible right* $K\Pi_n$-*module:* $I_{\geq n} = \bigoplus\limits_{l(P)=n} \langle P \rangle_K$ *is a decomposition into irreducible right* $K\Pi_n$-*modules.*

(viii) $I_{\geq n}$ *is an indecomposable left* $K\Pi_n$-*module.*

(ix) *Let* $(P_1, \cdots, P_k), (a_1, \cdots, a_n) \in \Pi_n$. *For every* $i \in \underline{k}$ *let* T_i *be the subset of* $\underline{n}$ *such that* $P_i = \{a_t \mid t \in T_i\}$ *and* φ_{T_i} *the unique monotone bijection between* $\underline{\mid T_i \mid}$ *and* T_i *with respect to the natural order* $\leq$ *on* $\mathbb{N}$. *Then,* $(P_1, \cdots, \overline{P_k}) \wedge (a_1, \cdots, a_n) = (a_{1\varphi_{T_1}}, \cdots, a_{\mid T_1 \mid \varphi_{T_1}}, \cdots, a_{1\varphi_{T_n}}, \cdots, a_{\mid T_n \mid \varphi_{T_n}})$ *is valid.*
Example: $(378, 16, 245) \wedge (3, 1, 8, 6, 4, 7, 5, 2) = (3, 8, 7, 1, 6, 4, 5, 2)$

(x) *The nilradical of the left* $K\Pi_n$-*module and of the algebra* $I_{\geq n}$ *are identical, and* $\langle e_{(1,2,\cdots,n-1,n)} \rangle_K$ *is a radical complement for the algebra* $I_{\geq n}$.

Proof: ad(i): This statement is a consequence of corollary 8.

ad(ii): For all $i \in \underline{n}$ by definition $e_{(i)} = (i)$ is valid, and thus, $e_{(1,2,\cdots,n-1,n)} = (1, 2, \cdots, n-1, n)$ is true, too. In addition, $K\Pi_n e_{(1,2,\cdots,n-1,n)}$ is contained in the ideal $I_{\geq n}$. We only need to prove now that all $P \in \Pi_n$ with $l(P) = n$ are contained in the left ideal $K\Pi_n e_{(1,2,\cdots,n-1,n)}$. Let $P \in \Pi_n$ with $l(P) = n$. From part (iii) we deduce $P = P \wedge (1, 2, \cdots, n-1, n)$.

ad(iii): Let $P, Q \in \Pi_n$ with $l(P) = n$. An element $\alpha \in S_n$ exists such that $P = (1\alpha, \cdots, n\alpha)$ is valid. $P \wedge Q = P$ is now a simple consequence of these facts. Hence, for all $a \in I_{\geq n}$ the right ideal $aK\Pi_n$ is one-dimensional. By using part (i) the set $I_{\geq n}$ is for $n \geq 2$ no right principal ideal. Based on theorem 9.4.1 in [10] we can deduce the statement about the uniseriality of $K\Pi_n$.

ad(iv): Based on corollary 1 every subalgebra of $K\Pi_n$ splits over K and is solvable. The group algebra KS_r is – by using theorem 3.2.20.1 in [22] – solvable if and only if the derived subgroup of S_r is a p-group and $char(K) = p$

54

is valid. For $r \geq 4$ is the alternating group A_r no p-group.

ad(v): Let $e_1 := (1, 2, 3)$, $e_2 := (1, 3, 2)$, $x := 1 + e_1$ and $y := 1 + e_1 - e_2$. We calculate:

$$
\begin{aligned}
& x^2 \\
=\ & (1 + e_1)^2 \\
=\ & 1 + 2e_1 + (e_1)^2 \\
=\ & 1 + 3e_1 \\
=\ & 1, \ (hence\ x\ is\ an\ involution)
\end{aligned}
$$

$$
\begin{aligned}
& y^2 \\
=\ & (1 + e_1 - e_2)^2 \\
=\ & 1 + e_1 - e_2 + e_1 + e_1 - e_1 - e_2 - e_2 + e_2 \\
=\ & 1 + 2e_1 - 2e_2 \\
=\ & 1 + e_2 - e_1,
\end{aligned}
$$

$$
\begin{aligned}
& y^3 \\
=\ & (1 + e_1 - e_2)(1 + e_2 - e_1) \\
=\ & 1 + e_2 - e_1 + e_1 + e_1 - e_1 - e_2 - e_2 + e_2 \\
=\ & 1 \ (thus,\ x\ is\ an\ element\ of\ order\ 3)\ and
\end{aligned}
$$

$$
\begin{aligned}
& xyx \\
=\ & (1 + e_1)(1 + e_1 - e_2)(1 + e_1) \\
=\ & (1 + e_1 - e_2 + e_1 + e_1 - e_1)(1 + e_1) \\
=\ & (1 - e_1 - e_2)(1 + e_1) \\
=\ & (1 + e_1 - e_1 - e_1 - e_2 - e_2) \\
=\ & 1 + e_2 - e_1 \\
=\ & y^2 \\
=\ & y^{-1} \ (x\ acts\ inverting\ by\ conjugation\ on\ y).
\end{aligned}
$$

The subgroup $\langle y, x \rangle_{E(K\Pi_n)}$ is a dihedral group of order 6 and isomorphic to S_3. We calculate $1 + y + y^2 = 1 + 1 + e_1 - e_2 + 1 + e_2 - e_1 = 3 = 0$.

ad(vi): The unitary one-dimensional K-algebras KS_1 and $I_{\geq 1} = K\Pi_1$ are isomorphic to K.

Theorem 1 lets us deduce that the ideal $I_{\geq 2}$ in $K\Pi_2$ is a local algebra possessing an one-dimensional nilradical. The radical complements of $K\Pi_2$ are – by using theorem 2 and example 1 – isomorphic to $K \times K$. The group algebra KS_2 is in the case $char(K) \neq 2$ semisimple and isomorphic to $K \times K$. In the other case[1] Maschke's theorem shows us that their nilradical is one-dimensional. Based on [14], chapter 1 three isomorphism-types of 2-dimensional associative unitary K-algebras exist: $K \times K$, a field extension of K or an algebra possessing a one-dimensional nilradical. Thus, the statement is true for $n = 2$.

Based on theorem 1 the ideal $I_{\geq 3}$ in $K\Pi_3$ is a local algebra. The group algebra KS_3 is not local because the group S_3 is no p-group (see theorem 1.1.21 in [23]). Thus, also the case $n = 3$ is proven.

For $n \geq 4$ the statement is true by using part (iv).

ad(vii): Based on part (iii) for every $P \in \Pi_n$ is the one-dimensional K-span $\langle P \rangle_K$ a right $K\Pi_n$-module (right principal ideal!) of $I_{\geq n}$, and thus it is irreducible. $I_{\geq n}$ is a direct sum of the irreducible and one-dimensional submodules $\langle P \rangle_K$ such that $l(P) = n$ is true.

ad(viii) and (x): Based on theorem 5.4 in [18] and (i) is $I_{\geq n} = K\Pi_n e_{(1,2,\cdots,n-1,n)}$ an indecomposable left $K\Pi_n$-module with module-radical $\langle e_{(1,2,\cdots,n-1,n)} - e_Q \mid Q \sim_{\Pi_n} (1,2,\cdots,n-1,n) \rangle_K$ of codimension 1. We use page 21 in [18] to deduce $e_P \wedge e_Q = e_P$ for all $P \sim_{\Pi_n} Q$. Thus, the module-radical is generated by nilpotent elements of nilpotency class 2. $K\Pi_n$ is solvable by using corollary 1, and hence proposition 5 in [21] shows us that the module-radical is contained in the nilradical of the algebra $I_{\geq n}$. $I_{\geq n}$ is not nilpotent, and thus, both radicals are identical. The one-dimensional K-span of the idempotent $e_{(1,2,\cdots,n-1,n)}$ is a radical complement.

ad(ix): A simple calculation let us verify this statement.$\diamond$

Definitions 6 Let A be a K-algebra and T a subset of A. We define $Ann_r(T) := \{a \in A \mid aT = \{0\}\}$ and $Ann_l(T) := \{a \in A \mid Ta = \{0\}\}$, and call $Ann_r(T)$ resp. $Ann_l(T)$ the right resp. left annulator (also annihilator) of T in A.$\diamond$

---

[1]Heinrich Maschke

Proposition 7 *Let K be a field and $n \in \mathbb{N}$. $K\Pi_n$ is only for $n = 1$ a (Quasi-)Frobenius algebra.*

Proof: Based on corollary 1 the field K is a splitting field for $K\Pi_n$. Thus, theorem 9.3.2 in [10] lets us deduce that $K\Pi_n$ is a Quasi-Frobenius algebra if and only if it is a Frobenius algebra. By definition is $K\Pi_n$ a Quasi-Frobenius algebra if for every right ideal R and for every left ideal L the identities $Ann_l(Ann_r(L)) = L$ and $Ann_r(Ann_l(R)) = R$ are valid. For $n = 1$ is $K\Pi_n$ isomorphic to the algebra[2] K which is a Frobenius algebra. Let $n \geq 2$. Let us focus on the right ideal $R := e_{(\underline{n})} K\Pi_n e_{(1,2,\cdots,n-1,n)}$ (see corollary 7.4 in [18]). Based on page 20 in [18] for all $P \in \Pi_n$ the equations $P \wedge e_{(\underline{n})} = 0$ (if $(\underline{n})$ is not smaller than P) and $P \wedge e_{(\underline{n})} = e_{P \wedge (\underline{n})}$ (if $(\underline{n})$ is smaller than P) are valid. The set partition $(\underline{n})$ is the 1-element in Π_n, and thus it is the only element of Π_n smaller than $(\underline{n})$. Hence, $Ann_l(R) = \langle P \in \Pi_n \mid l(P) \geq 2 \rangle_K (= I_{\geq 2})$ and $e_{(\underline{n})} \in Ann_r(Ann_l(R))$ are true. By using corollary 7.4 in [18] the set R is contained in the nilradical of $K\Pi_n$. The element $e_{(\underline{n})}$ is an idempotent of $K\Pi_n$ different from zero.$\diamond$

Proposition 8 *Let K be a field and $n \in \mathbb{N}$.*

   (i) *There is an ideal in $K\Pi_n$ different from the zero-space which is a principal ideal and a principal left and right ideal of $K\Pi_n$.*

   (ii) *There is an ideal in $K\Pi_n$ which is a principal ideal and a principal left ideal but for $n \geq 2$ no principal right ideal of $K\Pi_n$.*

   (iii) *Every principal left and right ideal of an associative unitary K-algebra is a principal ideal.*

   (iv) *There is an ideal in $K\Pi_n$ which is a principal ideal and a principal right ideal but for $n \geq 2$ no principal left ideal of $K\Pi_n$.*

   (v) *There is a left ideal in $K\Pi_n$ which is a principal left ideal but no principal ideal and no principal right ideal of $K\Pi_n$.*

   (vi) *There is a right ideal in $K\Pi_n$ which is a principal right ideal but for $n \geq 2$ no principal ideal and no principal left ideal of $K\Pi_n$.*

---

[2]Ferdinand Georg Frobenius

(vii) There is an ideal in $K\Pi_n$ that is for $n \geq 3$ no principal ideal and no principal left and right ideal of $K\Pi_n$.

(viii) Let B be a basis of $K\Pi_n$ and $a \in K\Pi_n$. Normally, there is no $b \in B$ such that $aK\Pi_n = bK\Pi_n$ is valid.

(ix) Let B be a basis of $K\Pi_n$ and $a \in K\Pi_n$. Normally, there is no $b \in B$ such that $K\Pi_n a = K\Pi_n b$ is true.

(x) Let B be a basis of $K\Pi_n$ and $a \in K\Pi_n$. Normally, there is no $b \in B$ such that $K\Pi_n a K\Pi_n = K\Pi_n b K\Pi_n$ is true.

Proof: ad(i)+(vii): By using corollary 4.6 in [18] is $e_{(\underline{n_l})} K\Pi_n e_{(1,2,\cdots,n-1,n)}$ a $(n-1)!$-dimensional ideal of $K\Pi_n$. Because of parts (ii)+(iii) of theorem 7 is $e_{(1,2,\cdots,n-1,n)} K\Pi_n = (1,2,\cdots,n-1,n) K\Pi_n$ one-dimensional and is K-spanned by the idempotent $(1,2,\cdots,n-1,n)$. The left ideal $K\Pi_n e_{(\underline{n_l})}$ is – based on page 22 in [18] – of dimension one and is K-spanned by the idempotent $e_{(\underline{n_l})}$. Let $a \in K\Pi_n$, $A := K\Pi_n$ and $x := e_{(\underline{n_l})} a(1,2,\cdots,n-1,n)$. Then, $AxA = xA = Ax = \langle x \rangle_K$ is valid, and we have proven parts (i) and (vii).

ad(ii): This statement is a direct consequence of part (iii) of theorem 7 with respect to the ideal $I_{\geq n}$.

ad(iii): Let A be an associative unitary K-algebra, $a, b \in A$ and $I := Aa = bA$. The identities $I = Aa \subseteq AaA \subseteq (bA)A \subseteq I$ and $bA \subseteq A(bA) \subseteq I$ are valid.

ad(iv): Let $I := e_{(\underline{n_l})} K\Pi_n$. Based on page 22 in [18] is the left ideal $K\Pi_n e_{(\underline{n_l})}$ of dimension one, and thus it is exactly the K-span of the idempotents $e_{(\underline{n_l})}$. Hence, $I = K\Pi_n e_{(\underline{n_l})} K\Pi_n$ is true. This ideal is – by using corollary 4.6 in [18] – of dimension $2 \cdot |\Pi_{n-1}| \geq 2$, and we have proven part (iv).

ad(v): Let $L := K\Pi_n e_{(\underline{n_l})}$. Based on page 22 in [18] the left ideal L if of dimension one, and thus it exactly the K-span of the idempotents $e_{(\underline{n_l})}$. In particular, $K\Pi_n e_{(\underline{n_l})} K\Pi_n = K\Pi_n e_{(\underline{n_l})}$ is valid. This ideal is – by using corollary 4.6 in [18] – of dimension $2 \cdot |\Pi_{n-1}| \geq 2$.

ad(vi): Let $R := (1, \cdots, n-1, n) K\Pi_n$. By using part (iii) of theorem 7 we deduce $R = \langle (1, \cdots, n-1, n) \rangle_K$. Thus, $K\Pi_n (1, \cdots, n-1, n) K\Pi_n = K\Pi_n (1, \cdots, n-1, n)$ is valid. This ideal is – based on part (ii) of theorem 7 – of dimension $n! \geq 2$.

ad(viii)-(x): Let $n = 2$, $A := K\Pi_2$, $char(K) = 2$ and $B := \Pi_n$. We determine all principal ideals, principal right and left ideals with respect to the basis B as generator:

principal ideals: $A(12)A = A, A(1,2)A = A(2,1)A = \langle(1,2),(2,1)\rangle$

principal right ideals: $(12)A = A, (1,2)A = \langle(1,2)\rangle_K, (2,1)A = \langle(2,1)\rangle_K$

principal left ideals: $A(12) = A, A(1,2) = A(2,1) = \langle(1,2),(2,1)\rangle_K$.

We calculate:

$((1,2) + (2,1))A = \langle(1,2) + (2,1)\rangle_K,$

$A((1,2) + (2,1)) = \langle(1,2) + (2,1), 2(1,2), 2(2,1)\rangle_K = \langle(1,2) + (2,1)\rangle_K$ and

$A((1,2) + (2,1))A = \langle(1,2) + (2,1)\rangle_K A = \langle(1,2) + (2,1)\rangle_K.$

This argumentation is used to prove the parts (ix) to (xi).⋄

Remark 12 Let A be an associative unitary K-algebra and $a, b \in A$ such that $I := aA = Ab$ is valid. Proposition 8 lets us deduce the identities $aA = AaA = Ab = AbA$. The ideal I is a principal ideal, too. An interesting question is whether a generator $s \in A$ exists such that $I = sA = As = AsA$ is valid simultaneously. We answer this question within the next chapter which might be viewed as an excursus.⋄

· Chapter 4 ·

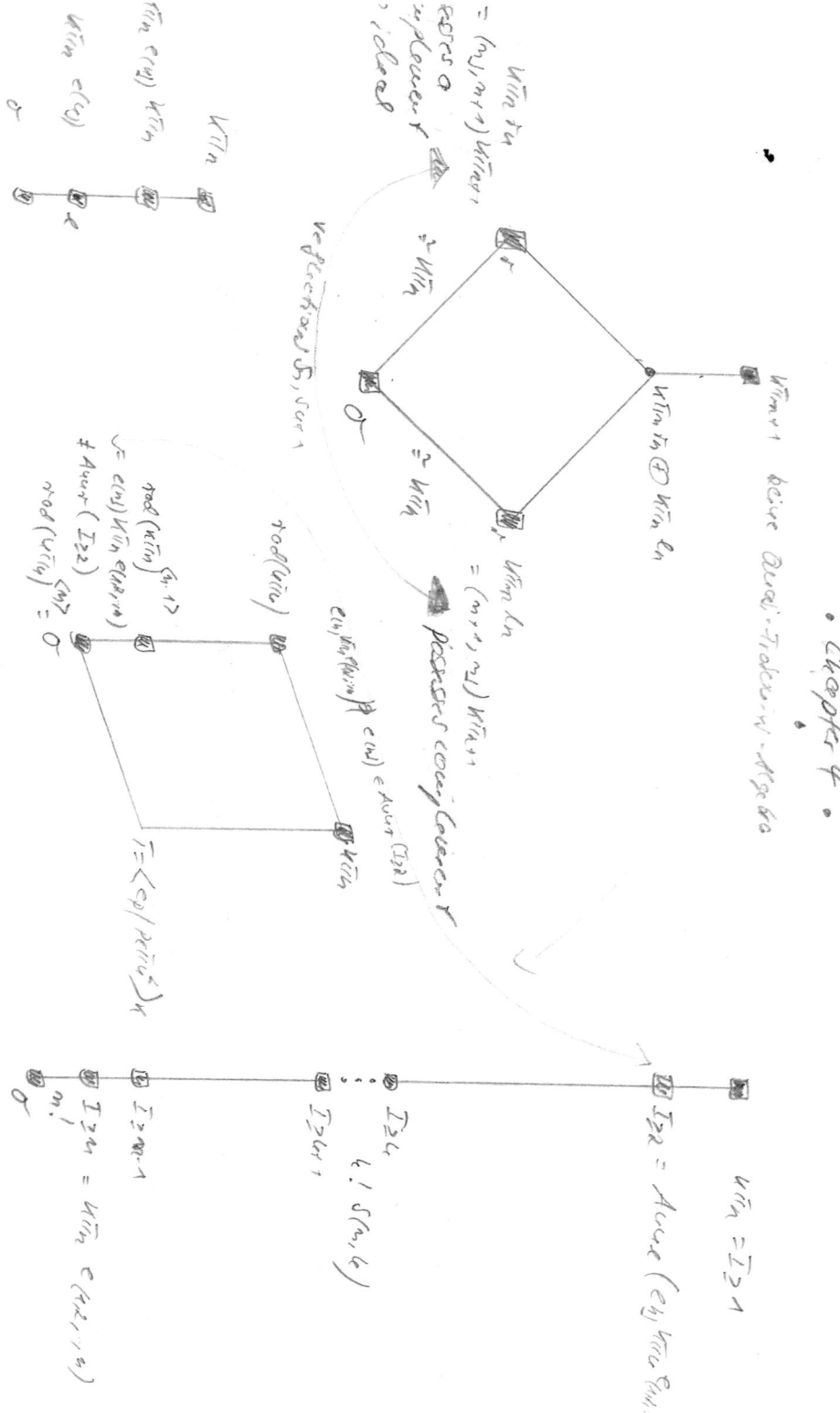

Reine Quasi-Fraktion-Algebra

— Over-sided and two-sided ideals of $k[[n]]$ —

4.3 Open-ended questions

- Let K be a field and $n \in \mathbb{N}$. Is the reflection s_n an inner or outer automorphism of $K\Pi_n$?

- Let K be a field and $n \in \mathbb{N}$. Determine all embeddings of $K\Pi_n$ in $K\Pi_{n+1}$!

- Let K be a field and $n \in \mathbb{N}$. For which $a \in K\Pi_n$ is $K\Pi_n \cdot a$ resp. $a \cdot K\Pi_n$ no right resp. left ideal?

- Let K be a field and $n \in \mathbb{N}$. Describe the lattice of all left and right ideals of $K\Pi_n$!

- Let K be a field and $n \in \mathbb{N}$. Calculate the dimension and a basis of all left and right ideals of $K\Pi_n$!

- Let K be a field and $n \in \mathbb{N}$. Describe the left and right annulator series of the nilradical of $K\Pi_n$ in $K\Pi_n$ resp. in the nilradical itself!

- Let K be a field and $n \in \mathbb{N}$. Does a subalgebra of $K\Pi_n$ exist which is isomorphic to KS_3?

- Let K be a field and $n \in \mathbb{N}$. Describe the left and right ideal lattice of $K\Pi_n$!

- Let K be a field and $n \in \mathbb{N}$. For which elements $a \in K\Pi_n$ is $K\Pi_n \cdot a$ resp. $a \cdot K\Pi_n$ a right resp. left ideal and thus identical to the principal ideal spanned by a?

4.4 Exercises

Exercise 44 *For $n = 1, 2, 3$ analyze the existence of a principal ideal which is not a principal right and left ideal in $K\Pi_n$. Is it possible to answer this exercise for an arbitrary natural number n?*

Exercise 45 *For all $P \in \Pi_3$ calculate the element Ps_3! For which $P \in \Pi_3$ is the identity $Ps_3 = P$ true?*

Exercise 46 *Let $n \in \mathbb{N}$. Prove $1_{\Pi_n} r_n = (\underline{n}) r_n = 1_{\Pi_n r_n} = (\underline{n}, n+1) \neq (\underline{n+1}) = 1_{\Pi_{n+1}}$ and $1_{\Pi_n} l_n = (\underline{n}) l_n = 1_{\Pi_n l_n} = (n+1, \underline{n}) \neq (\underline{n+1}) = 1_{\Pi_{n+1}}$.*

Exercise 47 *Determine the determinant of s_3 with respect to the basis Π_3 and $\{e_Q \mid Q \in \Pi_3\}$! Is it possible to answer this exercise for an arbitrary natural number n?*

Exercise 48 *Let $n \in \mathbb{N}$ and K be a field. Is $I_{\geq n}$ a right or left ideal?*

Exercise 49 *Let K be a field. Focus on the two bases Π_2, $\{e_P \mid P \in \Pi_2\}$ and Π_3, $\{e_P \mid P \in \Pi_3\}$ and determine their representative matrix with respect to the linear functions r_2, l_2 and s_2. Is it possible to answer this exercise for an arbitrary natural number n?*

Exercise 50 *For all $P \in \Pi_3$ describe Pl_3 and Pr_3. Does a connection between l_3 and r_3 exist considering s_2, s_3 and s_4?*

Exercise 51 *Let K be a field, $n \in \mathbb{N}$ and U, V two elements in $\{K\Pi_n s_n, K\Pi_n r_n, K\Pi_n l_n\}$ such that $U \neq V$ is valid. Determine the dimensions of U, V, $U \cap V$ and $U + V$! On what terms is the sum $U + V$ direct?*

Exercise 52 *Let K be a field. Is $K\Pi_4$ local, simple or a division algebra?*

Exercise 53 *Let K be a field. Is $K\Pi_3$ semisimple, commutative or separable?*

Exercise 54 *Let K be a field and $n \in \mathbb{N}$. Construct a left resp. right ideal of $K\Pi_n$ that is no right resp. left ideal and determine their dimensions.*

Exercise 55 *Is $I_{\geq 4}$ isomorphic to a group algebra? Do we need to know the connection to $K\Pi_n$ for a field K and a natural number n for answering this question?*

Exercise 56 *Let K be a field. Is $K\Pi_4$ a Quasi-Frobenius algebra? Is $K\Pi_5$ a Frobenius algebra?*

Exercise 57 *Let $P := (1,4,5,6,2,3)$ and $Q := (3, \{1,6,4\}, \{2,3\})$. Calculate PQ and QP! Are PQ, QP associated? Calculate PP, QQ, QPQ, PQP, $PQPQ$ and $QPQP$.*

Exercise 58 *Let $n, r \in \mathbb{N}$ and $P_1, \cdots, P_r \in \Pi_n$ such that $\mid \{P_1, \cdots, P_r\} \mid \leq 2$ is valid. Determine $P_1 P_2 \cdots P_r$?*

Exercise 59 *Let K be a field and $n \in \mathbb{N}$. What is the dimension of $I_{\geq n}$? Determine the dimension of $I_{\geq k}$ for all $k \in \underline{n}$! Calculate this dimension recursively by using theorem 6!*

Exercise 60 *Prove remark 11!*

Exercise 61 *Let $n \in \mathbb{N}$. Prove or disprove $l(PQ) = l(QP)$ for all $P, Q \in \Pi_n$! Are PQ, QP associated for all $P, Q \in \Pi_n$? Is $PQ = QP$ true for all $P, Q \in \Pi_n$?*

Exercise 62 *Determine elements $P, Q \in \Pi_6$ such that $\max\{6, l(P)l(Q)\} \neq l(PQ) \neq \max\{l(P), l(Q)\}$ is valid! What is the minimal Π_n such that such elements P, Q exist?*

Exercise 63 *Let K be a field. Do $K\Pi_3 r_3$ resp. $K\Pi_3 l_3$ possess a right ideal complement in $K\Pi_4$? What is the dimension of the summands in such a decomposition?*

Exercise 64 *Let K be a field and $n \in \mathbb{N}$. Is it true or false that $K\Pi_n l_n$, $K\Pi_n r_n$ and $K\Pi_n s_n$ are ideals, right or left ideals?*

Exercise 65 *Let K be a field, $n \in \mathbb{N}$ and $k, l \in \underline{n}$. Is it true or false that the set of ordered set partitions of length at least l for which one component is exactly $\{k\}$ are spanning an ideal, a right ideal, a left ideal or a subalgebra?*

Chapter 5

Duo algebras

Within this chapter we focus on the following topics related to Duo algebras initiated in remark 12:

- answering remark 12 in the context of bimodules

- consequences of the previous result

- characterizations of Duo algebras

- deduction of a necessary Lie property for Duo algebras and its application to $K\Pi_n$ and D_n.

5.1   A lemma of Tadashi Nakayama

We begin this section to answer the question raised in remark 12 which was already proven by Tadashi Nakayama[1] 1940 in [13] for finite-dimensional associative K-algebras. We generalize the answer within the context of cyclic bimodules.

Definitions 7 If V is a left A- and a right B-module and do the actions of A and B on V commute – that is $\forall v \in V, a \in A, b \in B : (av)b = a(vb)$ –, then we call V a (A, B)-bimodule. If A is a K-algebra, then we denote by A^- or by A^{op} the inverse or opposite algebra of A.. By $N(A)$ we symbolize the divisors of zero of A. An algebra-module is called unital if the 1-element

---

[1]Tadashi Nakayama

of the algebra acts as identity. A subalgebra of an algebra is called unital it
the 1-element of the algebra is contained in the subalgebra.⋄

Remark 13 Let A, B be algebras and V a (A, B)-bimodule. Then V is a
(B^{op}, A^{op})-bimodule. Because of $(A^{op})^{op} = A$ the opposite statement is true
as well.⋄

Lemma 3 *Let K be a field, A, B associative unitary K-algebras, V an unital
(A, B)-bimodule with respect to the left A-algebra action α and the right B-
algebra action β and $v, w \in V$ such that $Av = wB$ is valid and Av is finite-
dimensional. Then, $Av = vB = Aw = wB$ is true.*

Proof: First we prove $Av = vB$. Unital modules lets us deduce $v \in Av =
wB$. Thus, vB is a submodule of wB. By using the theorem of homo-
morphism for modules we derive $dim(Kern(v\beta)) \geq dim(Kern(w\beta))$. We
want to prove that $Kern(v\beta)$ is a submodule of $Kern(w\beta)$ and therefor
$Av = wB = vB$. Let $x \in Kern(v\beta)$. Then, $vx = 0$ is valid. Because of
$Av = wB$ an element $a \in A$ exists such that $w = av$ is true. By the definition
if bimodules we deduce $wx = (av)x = a(vx) = a \cdot 0 = 0$.
By using the opposite algebras A^{op} and B^{op} we conclude $vA^{op} = B^{op}w$, and
we use the facts proven so far to derive $B^{op}w = wA^{op}$. Thus, $wB = Aw$ is
valid.⋄

Corollary 10 *Let K be a field, A an associative unitary finite-dimensional
K-algebra, L, R two unital K-subalgebras of A and $v, w \in A$ such that $Lv =
wR$ is valid. Then, $Lv = vL = Rw = wR$ is true.*

Proof: This is a consequence of lemma 3 because A is an unital (L, R)-
bimodule with respect to the left and right unital algebra representation.⋄

We deduce the following corollary:

Corollary 11 *(Tadashi Nakayama, 1940) Let K be a field, A be an associative unitary finite-dimensional K-algebra and $v, w \in A$ such that $Av = wA$
is valid. Then, $Av = vA = Aw = wA$ is true.*⋄

The content of this corollary is that every principal left ideal (right ideal)
that is a principal right ideal (left ideal), too, then the same generator can
be used. Tadashi Nakayama has commented this fact in his article [13] to
be surprisingly and of special interest. We want to verify his comment by
deducing several corollaries. The reader may prove some more facts based
on this result within the exercises. Indeed, it is also a starting point for the
theory of Duo algebras.

Corollary 12 *Let K be a field, A be an associative unitary finite-dimensional K-algebra and $a \in A$. The following statements are valid:*

(i) a is left invertible if and only if a is right invertible.

(ii) a is a left divisor of zero if and only if a is a right divisor of zero.

(iii) $A = E(A) \,\dot\cup\, N(A)$

Proof: A is finite-dimensional, and thus – by the theorem of homomorphism – $Kern(a\rho)$ is the zero-space if and only if $Aa = A$ is valid. This statement is equivalent to the existence of an element $b \in A$ such that $ba = 1$ is true. Because of $Aa = A = 1 \cdot A$ and corollary 11 this fact is equivalent to $aA = A \cdot 1$. This is equivalent to the fact that a is left invertible. Again – by using the theorem of homomorphism – this statement is equivalent to $dim(Kern(a\lambda)) = 0$. Thus, the corollary is proven.$\diamond$

An easy consequence is the following corollary:

Corollary 13 *Let K be a field, A be an associative unitary finite-dimensional K-algebra and T an unital subalgebra of A. The following statements are valid:*

(i) $T = E(T) \,\dot\cup\, N(T)$

(ii) $E(T) = E(A) \cap T$

(iii) $N(T) = N(A) \cap T$.$\diamond$

Another consequence for Duo algebras is the following corollary:

Corollary 14 *Let K be a field and A be an associative unitary finite-dimensional K-algebra. The following statements are equivalent:*

(i) Every principal right ideal is a principal left ideal.

(ii) Every principal left ideal is a principal right ideal.

(iii) For all $x \in A$ the identity $Ax = xA$ is valid.

Proof: Let every principal right ideal a principal left ideal and $y \in A$. By our assumption an element $x \in A$ exists such that $Ax = yA$ is true. We use corollary 11 to deduce $yA = Ay = Ax = xA$. Thus, the left principal ideal Ay and the principal right ideal xA are identical.

Let every principal left ideal a principal right ideal and $y \in A$. Thus, an element $x \in A$ exists such that $Ax = yA$ is valid. Again by using corollary11 we derive $yA = Ay = Ax = xA$. In particular, $Ay = yA$ is valid.

Part (i) is a direct consequence of part (i), and thus the corollary is proven.$\diamond$

5.2 Characterization of Duo algebras

Definitions 8 Let A be a K-algebra. A is called a left Duo algebra resp. right Duo algebra if and only if every left ideal resp. every right ideal is an ideal. A is a Duo algebra if and only if A is a left and right Duo algebra.⋄

A direct consequence of these definitions is the following remark:

Remark 14 *Let A be an unitary K-algebra. The following statements are valid:*

(i) A ia a left Duo algebra if and only if for all $x \in A$ the identity $xA \subseteq Ax$ is valid.

(ii) A ia a right Duo algebra if and only if for all $x \in A$ the identity $Ax \subseteq xA$ is true.

(iii) A ia a Duo algebra if and only if for all $x \in A$ the identity $xA = Ax$ is valid.⋄

In [8] R.C. Courter has proven the following theorem which connects left and right Duo algebras:

Theorem 8 *(R.C. Courter, 1982) Let K be a field and A an associative finite-dimensional unitary K-algebra. The following statements are equivalent:*

(i) A is a left Duo algebra.

(ii) A is a right Duo algebra.

(iii) A is a Duo algebra.⋄

We can extend this theorems by using the results from the previous sections to characterize Duo algebras:

Main Theorem 1 *(Characterization of Duo algebras) Let K be a field and A an associative finite-dimensional unitary K-algebra. The following statements are equivalent:*

(i) A is a left Duo algebra.

(ii) A is a right Duo algebra.

(iii) A is a Duo algebra.

(iv) For all $x \in A$ the identity $xA \subseteq Ax$ is valid.

(v) For all $x \in A$ the identity $Ax \subseteq xA$ is valid.

(vi) For all $x \in A$ the identity $xA = Ax$ is valid.

(vii) Every principal right ideal is a principal left ideal.

(viii) Every principal left ideal a principal right ideal.

(ix) Every principal right ideal is an ideal.

(x) Every principal right ideal is a principal ideal.

(xi) Every principal left ideal is an ideal.

(xii) Every principal left ideal is a principal ideal.

(xiii) For all $x \in A$ the identity $xA = AxA$ is valid.

(xiv) For all $x \in A$ the identity $Ax = AxA$ is valid.

Proof: Parts (i)-(viii) are equivalent based on theorem 8, remark 14 and corollary 13.

The implications (xiii) $\to$ (x) $\to$ (ix) are straightforward to verify. Now let part (ix) be valid and let $x \in A$. Thus, the right ideal xA is an ideal containing the principal ideal AxA. This principal ideal contains xA, and therefor part (xiii) is valid. Hence, all parts (xiii), (x) and (ix) are equivalent. By using a similar argumentation we prove that all parts (xiv), (xii) and (xi) are equivalent.

Now we prove that the parts (xiii) and (xiv) are equivalent to (i)-(viii). Let (xiii) resp. (xiv) be valid and let $x \in A$. Then, $xA = AxA$ resp. $Ax = AxA$ is valid, and thus xA resp. Ax is an ideal. In particular, xA resp. Ax contains the left ideal Ax resp. the right ideal xA and part (v) is proven. Now let (vi) be valid and let $x \in A$. We derive $xA = Ax$, and therefor $AxA = xA = Ax$ is valid because $Ax = xA$ is an ideal containing xA and Ax.$\diamond$

A direct consequence of the parts (xiii) and (xiv) we derive the following corollary:

Corollary 15 *Let K be a field and A an associative finite-dimensional unitary K-Duo algebra. The following statements are valid:*

(i) Every principal ideal is a principal right ideal.

(ii) Every principal ideal is a principal left ideal.

(iii) Every principal ideal is a principal left ideal or a principal right ideal.⋄

Corollary 15 does not characterize Duo algebras which is illustrated by the following examples:

Example 3 Let K be a field and A a finite-dimensional associative simple K-algebra. A possesses only the trivial ideals A and the zero-subspace. In particular, A possesses only principal ideals: $A = A1A = 1A = A1$ and $0 = A0A = 0A = A0$. Thus, every principal ideal is a principal left ideal and a principal right ideal. But not every principal left ideal and principal right ideal are principal ideals: the so-called column-subspaces and row-subspaces are minimal right resp. left ideals of A which are principal right and left ideals generated by an idempotent of A.⋄

Example 4 Let K be a field of order 2. The following statements are valid:

(i) $K\Pi_2$ possesses exactly 8 elements.

(ii) $K\Pi_2$ is no Duo algebra.

(iii) Every principal ideal of $K\Pi_2$ is a principal left ideal or a principal right ideal.

The proof is to be done within exercise 71.⋄

Remark 15 Let A be an associative K-algebra. The principal ideals AxA of A are the cyclic submodules of A as left $A \otimes A^{op}$-module. We try to use the bimodule-lemma 3 with respect to the identity $AxA = yA$ where $A \otimes A^{op}$ acts from the left and A from the right side on the module A. But A is in general no bimodule with respect to this actions which is indirectly proven within example 4. Otherwise corollary 15 would characterize Duo algebras. In fact, both actions define a bimodule structure on A if and only if A is commutative.⋄

5.3 A necessary Lie condition

In this chapter we derive a necessary Lie condition for being a Duo algebra. Examples illustrate that this condition is not equivalent to the Duo property.

Lemma 4 *Let K be a field and A a finite-dimensional associative unitary solvable K-algebra for which K is a splitting field. If A is a Duo algebra, Then A° is a nilpotent K-Lie algebra.*

Proof: Based on a theorem of Salvatore Siciliano (see [19]) the Cartan subalgebras of A° are exactly the centralizers of the maximal tori of A°. Hence, A is Lie-nilpotent if and only if every maximal tori is central. Every separable element is contained in a tori by an enhanced version of the conjugacy part of the theorem of Wedderburn-Malcev: every separable subalgebra can be conjugated into a radical complement. The subalgebra generated by an separable element is separable. (for both statements see [22] or [29]) Thus, A is Lie-nilpotent if and only if every separable element is central. In [8] R.C. Courter has proven that every idempotent of a Duo algebra A is central. Let s be a separable element of A contained in a radical complement of A (again theorem of Wedderburn-Malcev, see [22]). We only need to prove that the radical complements are central. These complements are generated by idempotent elements because K is a splitting field for A and thus $A/rad(A)$ is isomorphic to K^n for some n. Idempotents are central and the lemma is proven.◇

Example 5 Let K be a field. We focus on the algebra A of strict lower triangular matrices over K in the case $n = 3$. Let a be the matrix for which every entry except $a_{2,1} = 1$ is zero. Then, $aA = 0$ is valid, and Aa is one-dimensional (For such matrices only the $a_{3,1}$-entry is different from zero). Hence, the nilpotent 3-dimensional associative K-algebra A is no Duo algebra.
Now we focus on (K, A) - the adjunction of an unit of A (see e.g. [22]). Within this algebra the identity $(0, a)(K, A) \neq (K, A)(0, a)$ is true. The 4-dimensional K-algebra (K, A) possesses a one-dimensional central radical complement. (K, A) Lie-nilpotent but it is not a Duo algebra.◇

Corollary 16 *Let K be a field and A a finite-dimensional associative unitary solvable K-algebra for which K is a splitting field and for which the radical complements are self-centralizing. If A is a Duo algebra, then A is semisimple.*

Proof: Based on lemma 4 the self-centralizing radical complements are central. Hence, A is a radical complement.◇

We present two examples for a Duo algebra:

Example 6 (1) Let K be a field and N be a zero K-algebra which is an algebra such that the multiplication is zero for every pair of elements. One example is to define such a zero multiplication on a K-space. Another example is the set of lower triangular matrices of $K^{2\times2}$. We proceed by the adjunction of an unit $A := (K; N)$ of N. Zero algebras are associative and commutative, and thus they are Duo algebras. Based on [22] the nilradical is exactly N and the radical factor algebra of $(K; N)$ isomorphic to K. In particular, $T := K \cdot 1$ is a central radical complement of $(K; N)$. We prove that such algebras are commutative, too. The radical centralizes itself because it is commutative, and it centralizes also T because T is central. Thus, it is centralizing the whole algebra $A = N \oplus T$. T is central, and hence $Z(A) = A$ is valid.

(2) Bell and Li analyzed group algebras. Let K be a field. They prove in [5] (2007) that the group algebra KQ_8 for the quaternion group is in the case $char(K) = 0$ a Duo algebra if and only if the equation $0 = 1 + x^2 + y^2$ has no solution in K for all $x, y \in K$. Thus, $\mathbb{Q}Q_8$ is a non-commutative Duo algebra. In addition, they prove for $char(K) = 2$ that KQ_8 is a Duo algebra if and only if the equation $0 = 1 + x + x^2$ has no solution in K. For example, $GF(2)Q_8$ is another non-commutative Duo algebra.◇

5.4   Duo-properties of $K\Pi_n$ and D_n

We deduce the conditions for $K\Pi_n$ and D_n being a Duo algebra.

Corollary 17 *Let K be a field and $n \in \mathbb{N}$. $K\Pi_n$ is only for $n = 1$ a Duo algebra.*

Proof: If $n = 1$ is valid, then $K\Pi_n$ is commutative and therefor a Duo algebra, too. Based on theorem 9 every radical complement of $K\Pi_n$ is self-centralizing. If $K\Pi_n$ is a Duo algebra, then $K\Pi_n$ is – by using corollary 16 – semisimple and therefor $n = 1$ based on theorem 1.◇

Corollary 18 *Let K be a field, $char(K) = 0$ and $n \in \mathbb{N}$. D_n is a Duo algebra if and only if $n \leq 2$ is valid.*

Proof: If $n \leq 2$, then D_n is commutative and therefor a Duo algebra. A theorem of Thorsten Bauer (see [3]) shows us that every radical complement of D_n is self-centralizing. If D_n is a Duo algebra, then D_n is semisimple based on corollary 16. M. D. Atkinson has proven within [2] that the nilpotency class of the nilradical of D_n is exactly $n - 1$. Thus, $n \leq 2$ must be valid.◇

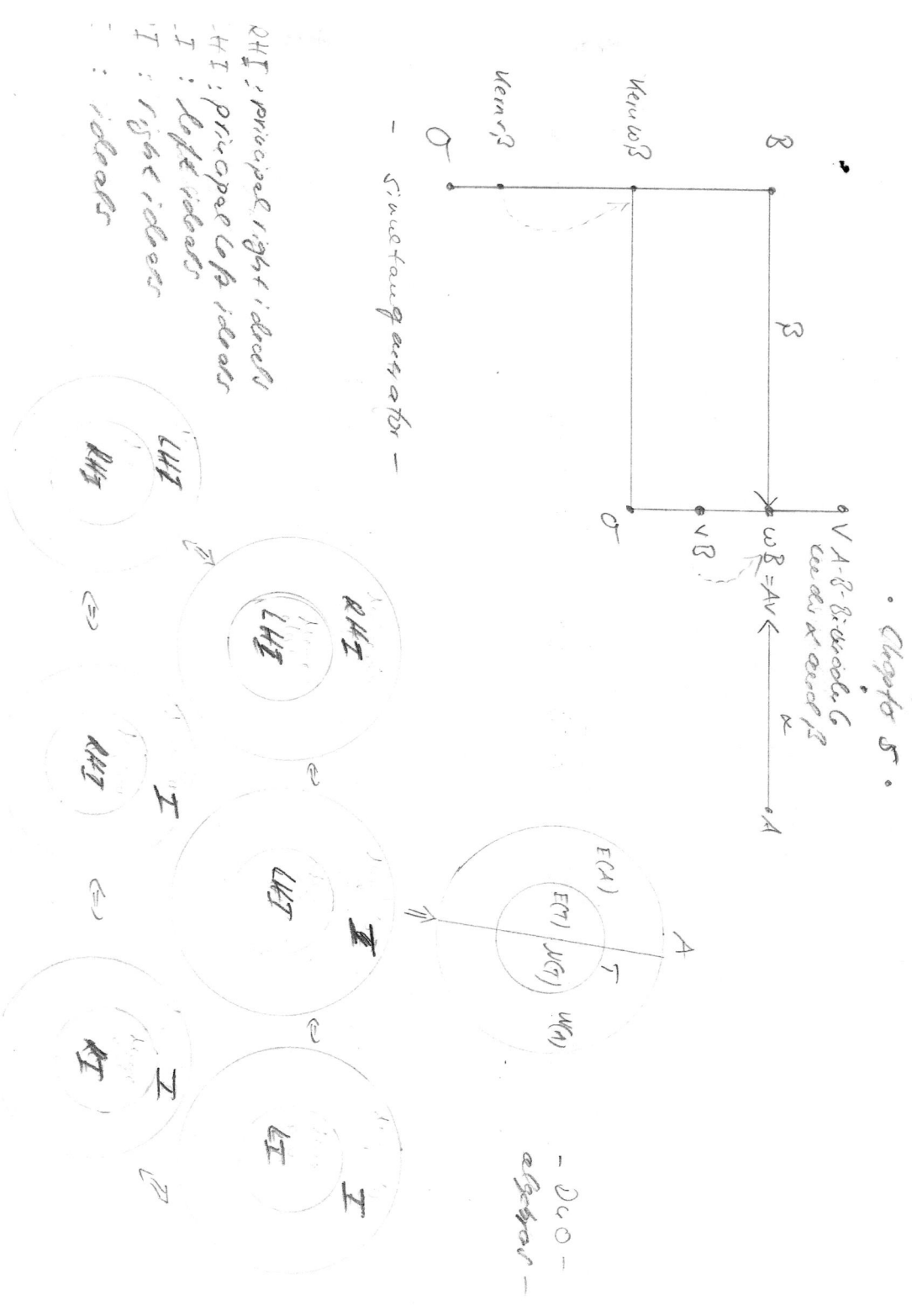

72

5.5 Open-ended questions

- Describe the structure of finite-dimensional associative unitary K-Duo algebras! Do certain classes within this class of algebras exist?

- Describe the structure of the algebras within corollary 15!

- Which Lie-nilpotent finite-dimensional associative unitary K-algebras are Duo algebras?

- Let A be an associative unitary K-algebra. How can we determine the elements $a \in A$ such that $aA = Aa$ is valid? Does an algebraic structure for the set of these elements exist?

5.6 Exercises

Exercise 66 *Prove lemma 11 by using the argumentation of lemma 3 and the left and right regular representation of a K-algebra!*

Exercise 67 *Let K be a field, A an associative unitary finite-dimensional K-algebra, T an unital K-subalgebra of A and $v, w \in A$ such that $Av = wT$ is valid. Prove $Av = vT = Aw = wT$.*

Exercise 68 *Let K be a field, A an associative unitary finite-dimensional K-algebra, T an unital K-subalgebra of A and $v, w \in A$ such that $Tv = wA$ is true. Prove the identity $Tv = vA = Tw = wA$.*

Exercise 69 *Let K be a field, A an associative unitary finite-dimensional K-algebra and $a \in \mathbb{N}$ such that $Aa = 0$ is valid. Prove the identity $aA = 0$.*

Exercise 70 *Let K be a field, $n \in \mathbb{N}$ and A, B elements of $K^{n \times n}$ such that $A \cdot B = 1$ is valid. Prove that A and B are inverse to each other.*[2]

Exercise 71 *Let K be a field of order 2. The following statements are valid:*

(i) $K\Pi_2$ possesses exactly 8 elements.

(ii) $K\Pi_2$ is no Duo algebra.

(iii) Every principal ideal of $K\Pi_2$ is a principal left or right ideal.

Exercise 72 *Prove the following statements:*

---

[2] By $M^{n \times m}$ we symbolize the set of $n \times m$-matrices over a set M.

(i) *If a diagonal matrix is invertible in $K^{n \times n}$, then its inverse is a diagonal matrix, too.*

(ii) *If an upper triangular matrix is invertible in $K^{n \times n}$, then its inverse is an upper triangular matrix, too.*

(iii) *If a lower triangular matrix is invertible in $K^{n \times n}$, then its inverse is a lower triangular matrix, too.*

(iv) *If A is a strict lower triangular matrix, then $1 + A$ is invertible in $K^{n \times n}$, and a strict lower triangular matrix B exists such that $1 + B$ is the inverse of $1 + A$.*

(v) *If A is a strict upper triangular matrix, then $1 + A$ is invertible in $K^{n \times n}$, and a strict upper triangular matrix B exists such that $1 + B$ is the inverse of $1 + A$.*

Exercise 73 *Let A be an unitary associative K-algebra. True or false:*

(i) *Every left divisor by zero is not right invertible and vice versa.*

(ii) *Every left divisor by zero is not left invertible and vice versa.*

(iii) *Every right divisor by zero is not right invertible and vice versa.*

(iv) *Every right divisor by zero is not left invertible and vice versa.*

(v) *Every left divisor by zero is a right divisor by zero and vice versa.*

(vi) *Every left invertible element is right invertible and vice versa.*

Exercise 74 *Prove corollaries 13 and 14!*

Exercise 75 *Let A be a K-algebra. For all $a, b \in A$ we define the relation $a \sim b :\Leftrightarrow aA = Ab$. Under the assumptions of Tadashi Nakayama's lemma 11 analyze the following statements:*

(i) *$\sim$ is left-total.*

(ii) *$\sim$ is reflexive.*

(iii) *$a \sim b$ for all a, b*

(iv) *$\sim$ is surjective.*

(v) *$\sim$ is symmetric.*

74

(vi) $\sim$ *is an equivalence relation.*

(vii) $\sim$ *is a congruence relation.*

(viii) $\sim$ *is transitive.*

(ix) $a \sim a$, $b \sim b$ *for all* a, b

(x) $a \sim b$, $c \sim d$ *for all* a, b, c, d

(xi) $ac \sim bd$ *for all* a, b, c, d

(xii) $a \sim c$, $b \sim c$ *for all* a, b, c

(xiii) $c \sim a$, $c \sim b$ *for all* a, b, c

(xiv) $\sim$ *is injective.*

(xv) $\sim$ *is bijective.*

(xvi) Every principal right ideal is a principal left ideal.

(xvii) Every principal left ideal is a principal right ideal.

Exercise 76 *Let A be an associative unitary K-algebra. We define $LR(A) := \{a \mid a \in A, aA = Aa\}$. Prove, disprove and analyze the following topics:*

(i) $LR(A)$ contains the center of A.

(ii) $LR(A)$ contains the group of units of A.

(iii) $LR(A)$ and A are identical.

(iv) A is a left Duo algebra if $LR(A) =$?

(v) A is a right Duo algebra if $LR(A) =$?

(vi) A is a Duo algebra if $LR(A) =$?

(vii) On what terms are $LR(A)$ and A identical?

(viii) $LR(A)$ is multiplicative closed.

(ix) $LR(A)$ is closed for scalar multiplication with K.

(x) $LR(A)$ is a K-subspace.

(xi) $LR(A)$ is a unital K-subalgebra.

(xii) *$LR(A)$ is a K-subspace if and only if it is a unital K-subalgebra.*

(xiii) *Let K be a field of order at least 3, A solvable and finite-dimensional. $LR(A)$ is a K-subspace if and only if it is identical to A. Thus, both statements are equivalent to the fact that A is a Duo algebra. (Tip: part (ii), calculate the K-span of the group of units of A!).*

Exercise 77 *Let K be a field and $n \in \underline{3}$. For which $e_Q, Q \in \Pi_n$ is the identity $e_Q K \Pi_n = K \Pi_n e_Q$ valid? Do you have a conjecture for an arbitrary $n \in \mathbb{N}$?*

Exercise 78 *Let K be a field and $n \in \underline{3}$. For which $Q \in \Pi_n$ is the identity $Q K \Pi_n = K \Pi_n Q$ true? Do you see a general rule for an arbitrary $n \in \mathbb{N}$?*

Exercise 79 *Let K be a field and $n \in \mathbb{N}$. Analyze whether the sets of lower and upper triangular matrices of $K^{n \times n}$ are Duo algebras.*

Exercise 80 *Let K be a field and A be a finite-dimensional associative unitary K-algebra. Prove that if A is a Duo algebra, then $A/rad(A)$ is a direct sum of finite many division algebras. From this result deduce that the set of nilpotent elements is an ideal which is exactly the nilradical. (Tip: Use the statement that all idempotents are central.)*

Chapter 6

Cartan subalgebras

Within this chapter we focus on the determination and description of Cartan subalgebras of the associated Lie algebra of $K\Pi_n$ and – in a more general context – of a finite-dimensional associative solvable K-algebra with splitting field K with respect to the following topics:

- connection between Pierce components and Cartan subalgebras

- description of self-centralizing radical components by special one-dimensional Pierce components

- descriptions of the algebra, its nilradical and its radical complements based on Pierce components

- $(K\Pi_n)^\circ$ possesses only self-centralizing radical components

- determination and description of the Cartan subalgebras of $K\Pi_n$.

6.1 Cartan subalgebras and Pierce components of associative solvable splitting K-algebras

Lemma 5 *Let K be a field, A an associative finite-dimensional K-algebra and $e_1, \cdots, e_n$ pairwise orthogonal idempotents of A such that $T := \langle e_1, \cdots, e_n \rangle_K$ is a complement of $\mathrm{rad}(A)$ in A. The following statements are valid:*

(i) K is a splitting field for A. In particular, A is solvable and $A/\mathrm{rad}(A)$ is separable.

(ii) $\sum_{i=1}^{n} e_i$ is the 1-element of T. In particular, T is unitary.

(iii) If A is unitary, then $1_A = 1_T$ is valid. In particular, T is unital.

(iv) $T \subseteq C_A(T)$

(v) If A is unitary, then $C_A(T) = \bigoplus_{i=1}^{n} e_i A e_i$ is true.

(vi) Let A be unitary. T is self-centralizing if and only if for all $i \in \underline{n}$ the Pierce component $e_i A e_i$ is contained in T.

(vii) Let A be unitary and T self-centralizing. The following statements are valid:

(a) $A = \bigoplus_{i,j=1}^{n} e_i A e_j$[1]

(b) $rad(A) = \bigoplus_{i \neq j=1}^{n} e_i A e_j$

(c) $T = \bigoplus_{i=1}^{n} e_i A e_i$

(d) $\forall i \in \underline{n} : e_i A e_i = \langle e_i \rangle_K$

(viii) Let A be unitary. T is self-centralizing if and only if for all $i \in \underline{n}$ the Pierce component $e_i A e_i$ is one-dimensional and identical to $\langle e_i \rangle_K$.

(ix) Let A be unitary. T is self-centralizing if and only if $T = \bigoplus_{i=1}^{n} e_i A e_i$ is true.

Proof: ad(i)+(ii): It is well-known that T and K^n are isomorphic and $\sum_{i=1}^{n} e_i$ is the 1-element of T.

ad(iii): This statement is a consequence of remark 1.10.1 in [22].

ad(iv): This part is a consequence of the commutativity of T (see (i)).

---

[1]Richard S. Pierce

ad(v): $e_1, \cdots, e_n$ are idempotent and pairwise orthogonal, and thus $\bigoplus_{i=1}^{n} e_i A e_i \leq$ $C_A(T)$ is true. Let $a \in C_A(T)$. Because of (iii) the identity $A = \bigoplus_{i,j=1}^{n} e_i A e_j$ is true, and we deduce for all $i, j \in \underline{n}$ the existence of elements $a_{i,j} \in A$ such that $a = \sum_{i,j=1}^{n} e_i a_{i,j} e_j$ is valid. Let $r \in \underline{n}$. The statements $a e_r = \sum_{i=1}^{n} e_i a_{i,r} e_r$ and $e_r a = \sum_{i=1}^{n} e_r a_{i,r} e_i$ are valid. Because of $a \circ e_r = 0$ and $A = \bigoplus_{i,j=1}^{n} e_i A e_j$ we deduce $e_i a_{i,r} e_r = 0 = e_r a_{r,i} e_i$.

ad(vi): This part is a consequence of parts (iv) and (v).

ad(vii): The part (a) is deductable by the two-sided Pierce decomposition and part (ii). The part (b) is a consequence of part (v), and part (d) can be concluded by part (c) and a dimension argument. For all $i \neq j \in \underline{n}$ the statement $(e_i A e_j)(e_i A e_j) = 0$ is true. Thus, the Pierce component $e_i A e_j$ is nilpotent. Because of the solvability of A (see part (i)) and proposition 5 in [21] the set $e_i A e_j$ is contained in the nilradical of A. Thus, part (vii) is deductable by the parts (a) and (c) and a dimension argument.

ad(viii): This part is a consequence of parts (v) and (vi).

ad(ix): This part is a consequence of parts (v) and (vi). $\diamond$

6.2 Cartan subalgebras and Pierce components of $K\Pi_n$

Definitions 9 Let K be a field, A a K-algebra and M a A-module with respect to the action δ. By $End_{(A;\delta)}(M)$ we denote the set of module-endomorphism of M with respect to A and δ. By λ resp. ρ we denote the left- resp. right-regular representation of A. $\diamond$

Lemma 6 *Let K be a field and $n \in \mathbb{N}$. For all $Q \in \Pi_n$ the endomorphism algebra $End_{(K\Pi_n e_Q;\rho)}(K\Pi_n e_Q)$ and the Pierce component $e_Q K\Pi_n e_Q$ are one-dimensional.*

Proof: Let $A := K\Pi_n$ and $e := e_Q$. The statement $eee = e \in eAe$ is valid. It is well-known that the function $\rho_{|eAe}$ is an algebra-isomorphism between eAe and $End_{(A;\rho)}(Ae)$. The latter endomorphism-algebra is – because of $Ae \leq A$

80

–contained in $End_{(Ae;\rho)}(Ae)$.

Let $\varphi \in End_{(Ae;\rho)}(Ae)$. By using theorem 5.4 in [18] the set $\{e_P \mid P \sim_{\Pi_n} Q\}$ is a K-basis of Ae. Let $e\varphi = \sum\limits_{P \sim_{\Pi_n} Q} k_P e_P$. Hence, for all $R \sim_{\Pi_n} Q$ is – based on page 21 in [18] – the equation $e_R e = e$ true. e is contained in Ae. and thus for all $R \sim_{\Pi_n} Q$ the calculation

$$
\begin{aligned}
& e_R \varphi \\
= & (e e_R)\varphi \\
= & (e\varphi)e_R \\
= & (\sum_{P \sim_{\Pi_n} Q} k_P e_P)e_R \\
= & (\sum_{P \sim_{\Pi_n} Q} k_P)e_R.
\end{aligned}
$$

is valid. Therefor, $\varphi = (\sum\limits_{P \sim_{\Pi_n} Q} k_P)id_{Ae}$ is true and the lemma is proven.$\diamond$

Corollary 19 *Let K be a field, $n \in \mathbb{N}$, $Q \in \Pi_n$ and $l(Q) \geq 2$.*

(i) $K\Pi_n e_Q$ is an indecomposable and not irreducible $K\Pi_n$-module by the action λ, and $End_{(K\Pi_n;\lambda)}(K\Pi_n e_Q)$ is isomorphic to K. In particular, it is a K-division algebra. (converse [2] of the lemma of Schur)

(ii) $K\Pi_n e_Q$ is a non-unitary associative algebra. In particular, the right-identity e_Q is no left-identity of $K\Pi_n e_Q$.

Proof: ad(i): Based on theorem 5.4 in [18] the set $K\Pi_n e_Q$ is an indecomposable but non-irreducible $K\Pi_n$-module under the action λ. Lemma 6 finishes the proof of part (i).

ad(ii): We assume that $K\Pi_n e_Q$ is an unitary algebra. Then $K\Pi_n e_Q$ would be an algebra isomorphic to $End_{(K\Pi_n e_Q;\rho)}(K\Pi_n e_Q)$. This endomorphism algebra possesses – based on lemma 6 – the dimension 1. But $K\Pi_n e_Q$ is – based on theorem 5.4 in [18] – of dimension $l(Q)! \geq 2$.$\diamond$

Theorem 9 *Let K be a field, $n \in \mathbb{N}$ and $T := \langle e_Q \mid Q \in \Pi_n^< \rangle_K$. The following statements are valid:*

---

[2]Issai Schur

(i) *The Cartan subalgebras of* $(K\Pi_n)^\circ$ *are the conjugates of* T *under* $1 + rad(K\Pi_n)$. *All radical complements are self-centralizing.*

(ii) $K\Pi_n = \displaystyle\bigoplus_{P,Q \in \Pi_n^{\leq}} e_P K\Pi_n e_Q$

(iii) $rad(A) = \displaystyle\bigoplus_{P \neq Q \in \Pi_n^{\leq}} e_P K\Pi_n e_Q$

(iv) $T = \displaystyle\bigoplus_{P \in \Pi_n^{\leq}} e_P K\Pi_n e_P$

(v) *For all* $P \in \Pi_n^{\leq}$ *the identity* $e_P K\Pi_n e_P = \langle e_P \rangle_K$ *is true.*

Proof: The result is a consequence of lemma 5, lemma 6, theorem 3 and theorem 5.10 in [3].$\diamond$

Corollary 1 *Let* K *be a field and* $n \in \mathbb{N}$. *The following statements are equivalent:*

(i) $(K\Pi_n)^\circ$ *is nilpotent.*

(ii) $K\Pi_n$ *is commutative.*

(iii) $n = 1$.

Proof: Based on corollary 9 the parts (ii) and (iii) are equivalent, and (i) is a direct consequence of (ii). Let $(K\Pi_n)^\circ$ be nilpotent. The Cartan subalgebras of $(K\Pi_n)^\circ$ are maximal nilpotent, and thus theorem 9 lets us deduce that $K\Pi_n$ is semisimple. Now we use corollary 9 to prove the commutativity of $K\Pi_n$.$\diamond$

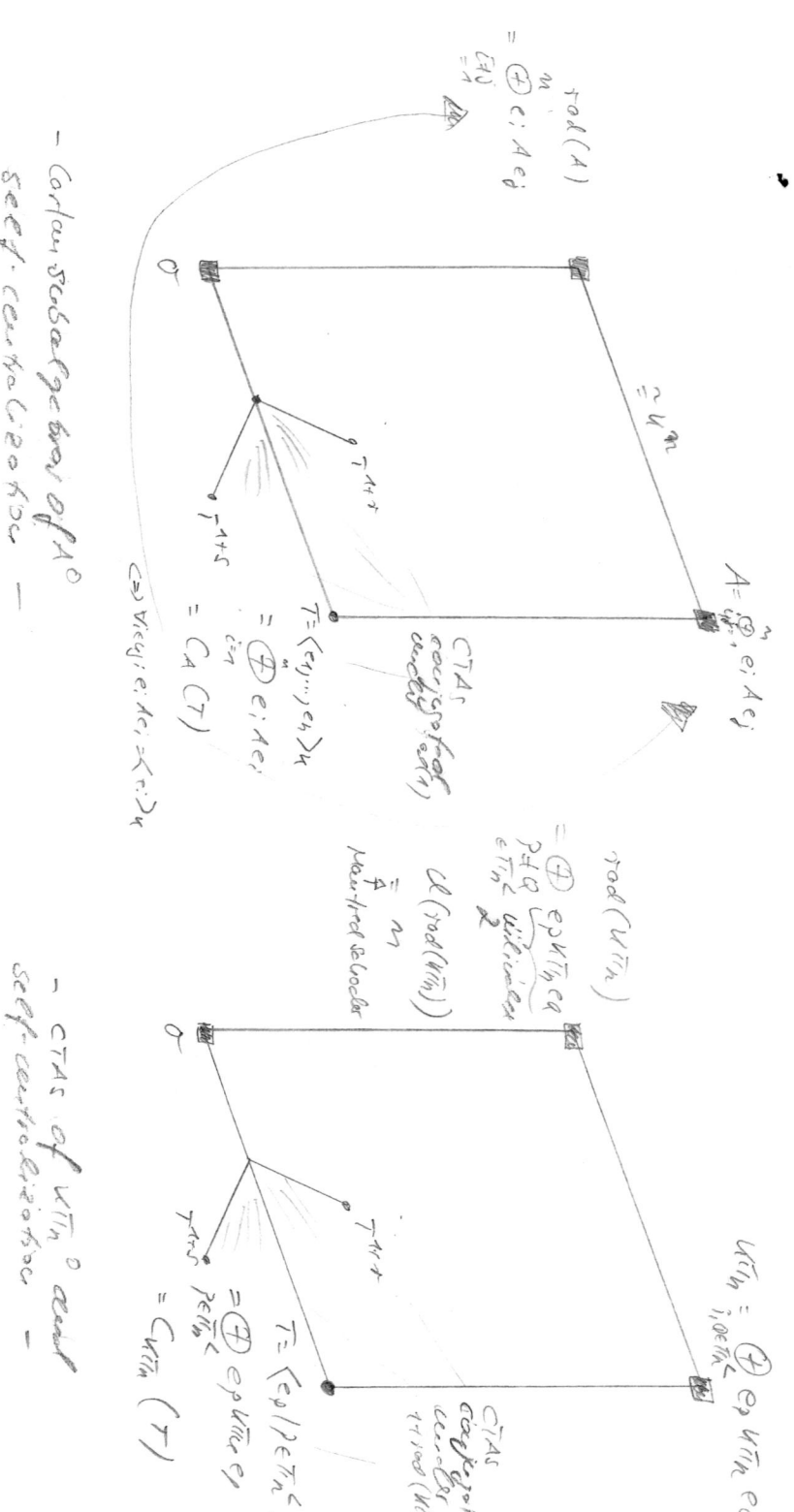

· Chapter 6 ·

6.3 Open-ended questions

- Let K be a field and M a finite idempotent monoid. Determine the Cartan subalgebras of the Lie algebra KM°! Which associative structure and which dimension do they have? Are they self-centralizing?

- How many Cartan subalgebras do KM for a finite field K and a finite idempotent monoid M possess?

6.4 Exercises

Exercise 81 *Let $n \in \mathbb{N}$. How many Cartan subalgebras do $K\Pi_n$ for a finite field K possess? (Tip: [22], [29])*

Exercise 82 *Let K be a field. Is $A := e_{(12,3)}K\Pi_3$ an unitary algebra, an indecomposable or irreducible right $K\Pi_3$-module?*

Exercise 83 *Let K be a field. Is $(12,3)K\Pi_3$ an unitary algebra, an indecomposable or irreducible right $K\Pi_3$-module?*

Exercise 84 *Let K be a field. Is $K\Pi_3(12,3)$ an unitary algebra, an indecomposable or irreducible left $K\Pi_3$-module?*

Exercise 85 *Let K be a field. Is $K\Pi_3 e_{(12,3)}$ an unitary algebra, an indecomposable or irreducible left $K\Pi_3$-module?*

Exercise 86 *Let K be a field. Calculate the dimensions of the Pierce components of $K\Pi_3$ with respect to the orthogonal idempotents $e_Q, Q \in \Pi_3^{\leq}$!*

Exercise 87 *Let K be a field. Determine at least two Cartan subalgebras of $K\Pi_3$!*

Exercise 88 *Let K be a field. Determine the dimensions of the Cartan subalgebras of $K\Pi_4$?*

Exercise 89 *Let K be a field. Is $(K\Pi_5)^\circ$ nilpotent?*

Exercise 90 *Let K be a finite field. How many Cartan subalgebras does $K\Pi_3$ possess?*

84

Exercise 91 *Let K be a field and $n \in \mathbb{N}$. Determine all Cartan subalgebras of the associated Lie algebra of the upper triangular matrices of $K^{n \times n}$. Are these Cartan subalgebras self-centralizing? Are the radical complements self-centralizing? Are the answers of these questions different for an algebraic closed field K?*

Exercise 92 *Let K be a field, A an associative finite-dimensional unitary solvable K-algebra possessing a separable radical factor algebra and let T be a radical complement in A. A result of T. Bauer lets us deduce that $C_A(T)$ is a Cartan subalgebra of A° (see [3]). By definition of the Cartan subalgebra is the subalgebra $C_A(T)$ identical to its Lie normalizer. Prove that the double-centralizer $C_A(C_A(T))$ is contained in the Cartan subalgebra. In addition, prove that T is contained in this double-centralizer. What are the answers in the case of a self-centralizing radical complement T?*

Chapter 7

Carter-, p'-Hall- and p-Sylow subgroups

Within this chapter we focus on the solvable group of units of $K\Pi_n$ and – in a more general context – of a finite-dimensional associative solvable K-algebra A for a finite field of characteristic p with respect to the following topics:

- description of the Carter subgroups of $E(K\Pi_n)$

- determination for the nilpotency and commutativity of $E(K\Pi_n)$

- determination of the p-Sylow subgroup of $E(A)$

- determination and counting of the p'-Hall subgroups of $E(A)$ and its connection to the Carter subgroups

- applications to $E(K\Pi_n)$.

7.1 Carter subgroups of $E(K\Pi_n)$

The next proposition is to be proven within the exercises (see also [22]):

Proposition 9 *Let K be a field and $n \in \mathbb{N}$. $E(K\Pi_n)$ is solvable: $E(K\Pi_n)$ is a semidirect product of the nilpotent normal subgroup $1 + rad(K\Pi_n)$ and the abelian subgroup $E(T)$ such that T is a radical complement.$\diamond$*

We use general results of T. Bauer to prove the following theorem:

Theorem 10 *Let K be a field, $n \in \mathbb{N}$ and $T := \langle e_Q \mid Q \in \Pi_n^\leq \rangle_K$. The following statements are valid:*

(i) In the case $\mid K \mid \neq 2$ the Carter subgroups of $E(K\Pi_n)$ are exactly the conjugates under $1 + rad(K\Pi_n)$ of $E(T) \cong E(K)^{|\Pi_n^{\leq}|}$.

(ii) In the case $\mid K \mid = 2$ the group $E(K\Pi_n)$ is nilpotent.

Proof: ad(i): This statement is a direct consequence of part (i) of theorem 9 and theorem 1 in [4].

ad(ii): Because of $\mid K \mid = 2$ the group of units of $K\Pi_n$ is exactly the nilpotent normal subgroup $1 + rad(K\Pi_n)$.$\diamond$

Corollary 2 Let K be a field and $n \in \mathbb{N}$. The following statements are valid:

(i) In the case $\mid K \mid \neq 2$ the group $E(K\Pi_n)$ is only for $n = 1$ nilpotent resp. abelian.

(ii) In the case $\mid K \mid = 2$ the nilpotent group $E(K\Pi_n)$ is only for $n \leq 2$ abelian.

Proof: ad(i): This part is a consequence of theorem 7 in [21] and corollary 1.

ad(ii): Let $J := rad(K\Pi_n)$. In this case $E(K\Pi_n) = 1 + J$ is true. Hence, $E(K\Pi_n)$ is abelian if and only if $J \circ J = 0$ is valid. Because of theorem 19 we deduce $J \circ J = J^{<2>}$. Thus, $K\Pi_n$ is abelian if and only if the class of nilpotency of J is not greater than 2. Based on corollary 7.4 in [18] this class is exactly n.$\diamond$

Example 7 Let K be a field and $n \in \mathbb{N}$. If the order of K is exactly 2, then the group of units of $K\Pi_n$ is nilpotent based on corollary 2 but the associated Lie algebra of $K\Pi_n$ has this property only for $n = 1$.
Let $n = 2$ and the order of K be exactly 2, $e_1 = (12), e_2 = (1,2)$ and $e_3 = (2,1)$. The sets $T_1 := \langle e_1, e_2 - e_1 \rangle_K$ and $T_2 := \langle e_1, e_3 - e_1 \rangle_K$ are two different radical complements which are the only ones.$\diamond$

7.2 p-Sylow- and p'-Hall subgroups in unit groups of solvable associative algebras

Proposition 10 Let K be a field, A a finite-dimensional associative unitary K-algebra, g an unit and T an unital K-subalgebra of A. The identity $E(T)^g = E(T^g)$ is valid.

Proof: We begin the proof by remarking that for every unital K-subalgebra S of A – based on corollary 13 – the identity $E(S) = S \cap E(A)$ is true. We deduce $E(T)^g = (T \cap E(A))^g = T^g \cap E(A)^g = T^g \cap E(A) = E(T^g)$.◇

Theorem 11 *Let K be a finite field of characteristic p, A a finite-dimensional associative unitary solvable K-algebra and T a radical complement. The following statements are valid:*

(i) $1 + rad(A)$ is the p-Sylow[1] subgroup of $E(A)$.

(ii) The p'-Hall subgroups are the conjugates of $E(T)$ under $1 + rad(A)$.

(iii) If K is a splitting field for A, then $E(T) \cong E(K)^{dim_K(A/rad(A))}$ is valid.

(iv) The centralizers of the p'-Hall subgroups are exactly the Carter subgroups of $E(A)$.

(v) If T is self-centralizing, then the p'-Hall subgroups are exactly the Carter subgroups of $E(A)$.

Proof: K is finite and therefor perfect. Thus, the radical factor algebra of A is separable, too. In particular, we can use the theorem of Wedderburn-Malcev to deduce the existence of a radical complement T. The algebra A is decomposed by $rad(A)$ and T as a direct sum. Based on corollary 1.1.8 in [23] the group $E(A)$ is a semidirect product of the normal subgroup $1 + rad(A)$ and the subgroup $E(T)$. A is solvable and therefor T is commutative. Thus, extension fields $K_1, \cdots, K_r$ of K exist such that T and $K_1 \times \cdots \times K_r$ are isomorphic. Again, based on corollary 1.1.8 in [23], we derive $E(T) \cong E(K_1) \times \cdots \times E(K_r)$. The theory of finite fields shows us that the order of each extension field K_i $(i \in \underline{r})$ is a p-power. Because of $E(K_i) = K_i \setminus \{0\}$ is $\mid E(K_i) \mid$ – for all $i \in \underline{r}$ – of p'-order. $rad(A)$ is a K-space, and thus part (i), (iii) and the first part of (ii) are valid.

Let H a p'-Hall subgroup of $E(A)$. By using a theorem of Hall an element $g \in E(A)$ exists such that $H = E(T)^g$ is valid. T and its conjugate T^g are radical complements, and thus by the theorem of Wedderburn and Malcev and $r \in rad(A)$ exists such that $T^g = T^{1+r}$ is valid. We use proposition 10 to conclude $H = E(T)^g = E(T^g) = E(T^{1+r}) = E(T)^{1+r}$

---

[1]Peter Ludwig Mejdell Sylow

The statements in (iv) and (v) are a consequence of (ii) and theorem 1 in [3].⋄

Proposition 11 *Let K be a field, A a finite-dimensional associative unitary solvable K-algebra and T a separable radical complement. The identity $N_{E(A)}(E(T)) = C_{E(A)}(E(T))$ is valid. In the case $\mid K \mid \neq 2$ the equation $C_{E(A)}(E(T)) = E(C_A(T))$ is true.*

Proof: Let $g \in N_{E(A)}(E(T))$. The algebra A is decomposed by $rad(A)$ and T as a direct sum. We use corollary 1.1.8 in [23] to deduce that $E(A)$ is a semidirect product of the normal subgroup $1 + rad(A)$ and the subgroup $E(T)$. Thus, elements $r \in rad(A)$ and $t \in E(T)$ exist such that $g = t(1 + r)$ is valid. By using the commutativity of T we deduce $E(T) = E(T)^g = E(T)^{t(1+r)} = E(T)^{1+r}$, and thus $E(T)(1 + r) = (1 + r)E(T)$ is true. Let $s \in E(T)$. An element $x \in E(T)$ exists such that $s(1+r) = (1+r)x$ is valid. We conclude $s + sr = x + rx$, and by using $sr, rx \in rad(A)$ and $s, x \in E(T)$ we deduce $s = x$ and $sr = rs$. Thus, r centralizes every element of $E(T)$. Therefor, $g \in C_{E(A)}(E(T))$ is proven.

In the case $\mid K \mid \neq 2$, we use lemma 5.17 in [3] to derive $T = \langle E(T) \rangle_K$, and the add-on is proven.⋄

Theorem 12 *Let K be a finite field of characteristic p, A a finite-dimensional associative unitary solvable K-algebra and T a radical complement. The following statements are valid:*

(i) *There are exactly $\frac{|E(A)|}{|C_{E(A)}(E(T))|}$ p'-Hall subgroups in $E(A)$. In particular, their number is a divisor of the order of the p-Sylow subgroup $1 + rad(A)$ of $E(A)$.*

(ii) *The number of p'-Hall subgroups is exactly the number of the Carter subgroups of $E(A)$.*

(iii) *If T self-centralizing and $\mid K \mid \neq 2$, then exactly $\mid rad(A) \mid$ p'-Hall subgroups exist within $E(A)$. (This is the maximal possible number of p'-Hall subgroups of $E(A)$.)*

Proof: ad(i): This part is a consequence of the orbit-stabilizer theorem and proposition 11.

ad(ii): Carter subgroups are self-normalizing. We use part (i) and theorem 11 to deduce part (ii).

ad(iii): This part is a consequence of part (i) and proposition 11. The add-on is deductable by the commutativity of T.⋄

7.3 Consequences for the group of units of $K\Pi_n$

Corollary 3 *Let K be a finite field of characteristic p, $n \in \mathbb{N}$ and T a radical complement in $K\Pi_n$. The following statements are valid:*

(i) $1 + rad(K\Pi_n)$ *is the p-Sylow subgroup of $E(K\Pi_n)$.*

(ii) *The p-Sylow subgroup is of order $\mid K \mid^{|\Pi_n|-B(n)}$.*

(iii) *The p'-Hall subgroups are the conjugates of $E(T)$ under $1 + rad(K\Pi_n)$.*

(iv) $E(T) \cong E(K)^{B(n)}$

(v) *The p'-Hall subgroups are the exactly the Carter subgroups.*

(vi) *In the case $\mid K \mid \neq 2$ exactly $\mid K \mid^{\sum\limits_{k=0}^{n} (k!-1)\, S(n,k)}$ p'-Hall subgroups exist in $E(K\Pi_n)$.*

(vii) *In the case $\mid K \mid = 2$ exactly one p'-Hall subgroup exists in $E(K\Pi_n)$.*

Proof: This corollary is a direct consequence of theorems 12 and 11 and corollary 1.⋄

7.4 Open-ended questions

- Let K be a field and M a finite idempotent monoid. On what terms is $E(KM)$ abelian or nilpotent?

- Let K be a field and M a finite idempotent monoid. What are the Carter subgroups of $E(KM)$?

- Let K be a field and M a finite idempotent monoid. On what terms is $E(KM)$ solvable?

- Is theorem 11 true without assuming the solvability?

- Is proposition 11 true without assuming the solvability?

- Is theorem 12 true without assuming the solvability?

- Let K be a finite field of characteristic p. What are the q-Sylow subgroups of $E(A)$ for a finite-dimensional unitary associative K-algebra A ($q \neq p$ another prime number)? What is the answer for $A = KM$ for a finite idempotent monoid M and what for $A = K\Pi_n$ for $n \in \mathbb{N}$? What is the number of Carter subgroups of $E(A)$? What is the normalizer of a Carter subgroup of $E(A)$?

7.5   Exercises

Exercise 93 *Let K be a field. Determine all Carter subgroups of $E(K\Pi_3)$!*

Exercise 94 *Let K be a field. Is $E(K\Pi_4)$ nilpotent?*

Exercise 95 *Let K be a field. Is $E(K\Pi_5)$ abelian?*

Exercise 96 *Let K be a field and $n \in \mathbb{N}$. Prove that $E(K\Pi_n)$ is solvable.*

Exercise 97 *Prove proposition 9!*

Exercise 98 *Let K be a finite field of characteristic p. How many p-Sylow subgroups does $E(K\Pi_9)$ possess?*

Exercise 99 *Let K be a finite field of characteristic p. How many p'-Hall subgroups does $E(K\Pi_4)$ possess?*

Exercise 100 *Let K be a finite field of characteristic p. Describe the Carter subgroups of $E(K\Pi_4)$ up to isomorphism.*

Exercise 101 *Let K be a finite field of characteristic p. Determine the class of nilpotency of the p-Sylow subgroups of $E(K\Pi_3)$.*

Exercise 102 *Let K be a field, $n \in \mathbb{N}$ and A the algebra of lower triangular matrices of $K^{n \times n}$. True or false:*

- *A is isomorphic to the algebra of upper triangular matrices of $K^{n \times n}$.*

- *A is anti-isomorphic to the algebra of upper triangular matrices of $K^{n \times n}$.*

- *A is anti-isomorphic to the algebra of lower triangular matrices of $K^{n \times n}$.*

- *The nilradical of A is the set of strict lower triangular matrices.*

- *A is solvable and the set of diagonal matrices id a self-centralizing radical complement.*

- $E(A)$ *is solvable.*

- *Determine the Carter subgroups of $E(A)$?*

- *What are the consequences of theorems 11 and 12 for $E(A)$ if K is finite?*

Exercise 103 *Transfer exercise 92 to Carter subgroups.*

Chapter 8

The center

Within this chapter we focus on the center of $K\Pi_n$ and its group of units:

- determination of the center of $K\Pi_n$

- description of the center of $K\Pi_n$ by intersecting subalgebras

- conditions for $K\Pi_n$ being direct-indecomposable

- internal description of the center of Π_n by using classes with respect to $\sim$

- internal description of the center of the group of units of $K\Pi_n$

- external description of the center of the group of units of $K\Pi_n$ by using Carter subgroups

- connection between the center of the group of units and the group of units of the center of $K\Pi_n$.

8.1 The center of $K\Pi_n$

Theorem 13 *Let K be a field and $n \in \mathbb{N}$. $K\Pi_n$ is central.*[1]

<u>Proof:</u> The center of $K\Pi_n$ is contained in every maximal nilpotent subalgebra of $(K\Pi_n)^\circ$. Hence, $Z(K\Pi_n)$ is contained in the Cartan subalgebra (see theorem 9) $\langle e_Q \mid Q \in \Pi_n^{\leq} \rangle_K$. Let $z \in Z(K\Pi_n)$ like $z = \sum\limits_{Q \in \Pi_n^{\leq}} k_Q e_Q$. Let $P, Q \in \Pi_n^{\leq}$ such that $Q < P$ is valid. Based on theorem 6.4 in [18] we derive $dim_K(e_P K\Pi_n e_Q) \geq 1$. Let $a \in K\Pi_n$ such that $e_P a e_Q \neq 0$ is true. The

---

[1]An unitary K-algebra is called central if its center is exactly $K \cdot 1_A$.

93

idempotents $e_R, R \in \Pi_n^{\leq}$ are – based on theorem 3 – pairwise orthogonal. We calculate:

$$
\begin{aligned}
0 \\
&= z \circ (e_P a e_Q) \\
&= \sum_{R \in \Pi_n^{\leq}} k_R(e_R \circ (e_P a e_Q)) \\
&= k_P e_P a e_Q - k_Q e_P a e_Q \\
&= (k_P - k_Q) e_P a e_Q.
\end{aligned}
$$

For all $P, Q \in \Pi_n^{\leq}$ such that $Q < P$ is valid we have proven $k_P = k_Q$. Let $Q := (1, 2, \cdots, n)$. We use theorem 7 to deduce for all $P \in \Pi_n$ the statement $Q \wedge P = Q$. Hence, $Q < P$ is true. The identity $min\{1\} < min\{2\} < \cdots < min\{n\}$ is valid, and thus we conclude $Q \in \Pi_n^{\leq}$ and $z = k_{(1,2,\cdots,n)} \sum_{P \in \Pi_n^{\leq}} e_P$. The sum of the idempotents $e_R, R \in \Pi_n^{\leq}$ is – based on theorem 3 – exactly the 1-element of their K-span. This span is – based on the same theorem – a radical complement. We use lemma 5 to deduce that the sum is the 1-element $K\Pi_n$. Thus, we have proven $z = k_{(1,2,\cdots,n)} 1_{K\Pi_n}$. $\diamond$

Corollary 4 *Let K be a field and $n \in \mathbb{N}$. The following statements are valid:*

(i) *The center of $K\Pi_n$ is the intersection of all radical complements of $K\Pi_n$.*

(ii) *The center of $K\Pi_n$ is the intersection of all Cartan subalgebras of $(K\Pi_n)^\circ$.*

(iii) *The center of $K\Pi_n$ is the intersection of all maximal nilpotent subalgebras of $(K\Pi_n)^\circ$.*

(iv) *$K\Pi_n$ is direct-indecomposable.*

Proof: ad(i): $K\Pi_n$ is solvable, and thus the intersection of all radical complements is exactly the radical complement of the center of $K\Pi_n$ (see corollary 5.1.5 in [22]). We use theorem 13 to deduce the semi-simplicity of the center. Thus, part (i) is proven.

ad(ii): Based on theorem 9 the Cartan subalgebras of $(K\Pi_n)^\circ$ are exactly the radical complements of $K\Pi_n$. By using part (i) we deduce part (ii).

ad(iii): The center is contained in every maximal nilpotent subalgebra of

$(K\Pi_n)^\circ$. Thus, part (iii) is deductable by part (ii): Cartan subalgebras are maximal nilpotent.

ad(iv): This part is a direct consequence of theorem 13.$\diamond$

8.2 The center of Π_n

Proposition 12 *Let $n \in \mathbb{N}$, s_n the reflection on Π_n (see proposition 3) and K a field. The following statements are valid:*

(i) For all $P \in \Pi_n$ the identity $P \sim_{\Pi_n} Ps_n$ is valid.

(ii) s_n possesses only one fix point: 1_{Π_n}.

(iii) $s_r, r \in \mathbb{N}$ are inducing anti-automorphism of order 2 on $(K\Pi; \vee)$.

Proof: ad(i)+(ii): Let $P := (P_1, \cdots, P_k) \in \Pi_n$. In the case $k \geq 2$ the identity $P \neq Ps_n = (P_k, \cdots, P_1)$ is valid. In addition, $(P_1, \cdots, P_k) \wedge (P_k, \cdots, P_1) = (P_1, \cdots, P_k)$ and $(P_k, \cdots, P_1) \wedge (P_1, \cdots, P_k) = (P_k, \cdots, P_1)$ are true. Thus, we have proven parts (i) and (ii).

ad(iii): Let $P := (P_1, \cdots, P_k)$ and $Q := (Q_1, \cdots, Q_l)$ two elements of Π_n such that $\bigcup\limits_{i=1}^{l} P_i \neq \bigcup\limits_{i=1}^{k} Q_i$ is valid. We calculate:

$$
\begin{aligned}
& (P \vee Q)s_{k+l} \\
= & (P_1, \cdots, P_k, Q_1, \cdots, Q_l)s_{k+l} \\
= & (Q_l, \cdots, Q_1, P_k, \cdots, P_1) \\
= & Qs_l \vee Ps_l.
\end{aligned}
$$

Thus, we have finished the proof.$\diamond$

Corollary 5 *Let $n \in \mathbb{N}$. $Z(\Pi_n) = \{1\}$ is valid, and the center of Π_n consists exactly of the classes of order 1 of Π_n with respect to $\sim_{\Pi_n}$.*

Proof: Based on remark 3 is every class of Π_n with respect to $\sim_{\Pi_n}$ which contains a central element of order one. Let $Z \in \Pi_n$ such that $[Z]_{\sim_{\Pi_n}} = \{Z\}$ is valid. Proposition 12 lets us deduce that $Z \sim_{\Pi_n} Zs_n$ and $Z = Zs_n$ are true. The same proposition yields to $Z = 1$. The second part is a direct consequence of theorem 13.$\diamond$

96

Remark 16 Within proposition 12 we have defined an anti-automorphism of order 2: an involution. This fact is true within several contexts in algebra: the transpose of a matrix, the inverse of a group element, the conjugate within quaternion algebras. Thus, the question arises if the existence of a non-trivial anti-automorphism induces the existence of an involution. A positive answer is given by[2] Adrian Albert in [1] for central-simple associative finite-dimensional algebras. If we focus on[3] a Galois extension of $\mathbb{Q}$ of order 3, then within this extension no involution exists: every anti-automorphism is an automorphism but 2 is no divisor of the Galois group. Scharlau constructs in [16] an example of an associative non-commutative algebra for which the answer is negative. Morandi, Sethuramam and Tagnol construct in [12] another example of this kind but here an anti-automorphism is no linear functions.$\diamond$

8.3 The center of the group of units of $K\Pi_n$

Proposition 13 *Let A be a finite-dimensional associative unitary K-algebra. The following statements are valid:*

(i) If S, T are unital K-subalgebras of A, then $E(T \cap S) = E(T) \cap E(S)$ is true.

(ii) Let A be solvable, $\mid K \mid \neq 2$, $A/rad(A)$ separable and T be a radical complement. Then $A = \langle E(A) \rangle_K$, $E(Z(A)) = Z(E(A))$ and $E(C_A(T)) = C_{E(A)}(E(T))$ are valid. In particular, A is an epimorphic image of the group algebra $K(E(A))$.

Proof: ad(i): Based on lemma 5.6 in [22] the identities $E(T) = E(A) \cap T$, $E(S) = E(A) \cap S$ and $E(T \cap S) = E(A) \cap T \cap S$ are valid, and we have proven part (i).

ad(ii): We use remark 1 in [4] to deduce $T = \langle E(T) \rangle_K$. Thus, $E(C_A(T)) =$

[2]Abraham Adrian Albert

[3]Évariste Galois

$C_{E(A)}(E(T))$ is valid. Because of $A = rad(A) \oplus T$ and corollary 1.1.8 in [23] we derive that $E(A)$ is the semidirect product of $1 + rad(A)$ and $E(T)$. We conclude $\langle E(A) \rangle_K = A$ and $E(Z(A)) = Z(E(A))$.◇

Corollary 6 *Let K be a field, $n \in \mathbb{N}$ and $A := K\Pi_n$.*

(i) *The intersection of all unit groups of all radical complements of A is exactly the group of units of the center of A.*

(ii) *For $| K | \neq 2$ the intersection of all unit groups of all radical complements of A is exactly the center of the group of units of A.*

(iii) *The intersection of all Carter subgroups of $E(A)$ is exactly the group of units of the center of A.*

(iv) *In the case $| K | \neq 2$ is the intersection of all Carter subgroups of $E(A)$ is exactly the center of the group of units of A.*

(v) *In the case $| K | \neq 2$ the center of the group of units of A is exactly $E(K \cdot 1) = (K \setminus \{0\}) \cdot 1$.*

(vi) *In the case $| K | = 2$ the center of the group of nits of A is exactly the center of $1 + rad(A)$.*

Proof: Based on theorem 2 the algebra A is solvable and $A/rad(A)$ separable. We use corollary 4, theorem 13 and proposition 13 to finish the proof.◇

∘ Chapter 7+8 ∘

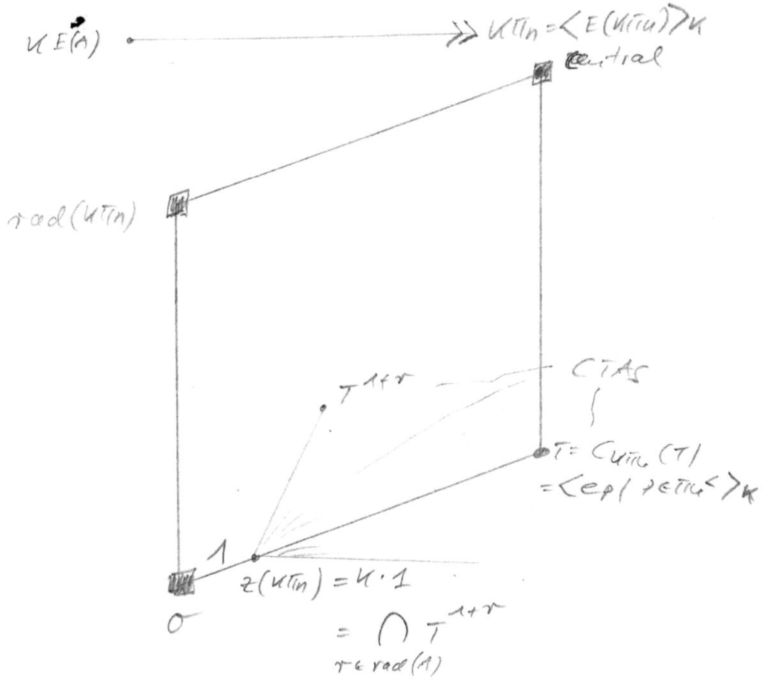

$\mathcal{U} \in (\Lambda)$ ⟹ $\mathcal{U}(K\bar\tau_m) = \langle E(K\bar\tau_m)\rangle_K$ central

$\uparrow \mathrm{rad}(K\bar\tau_m)$

$\mathrm{rad}(K\bar\tau_m)$

0

1

$Z(K\bar\tau_m) = K\cdot 1$
$= \bigcap_{\tau \in \mathrm{rad}(\mathcal{U})} \tau^{-1+\tau}$

$\tau^{-1+\tau}$

$F = \mathcal{U}_{K\bar\tau_m}(\tau) \subset \tau$
$\{ = \langle e_p | p \in \tau_m \rangle_K$

$\subset \tau \subset \mathcal{U}$

$\uparrow + \mathrm{rad}(\mathcal{U}\bar\tau_m)$

$\mathcal{U}$ p-Sylow $|\mathcal{U}| = |\tau_m| \cdot |\tau_m|$

$Z(E(K\bar\tau_m)) = E(Z(K\bar\tau_m))$
$= \bigcap_{\tau \in \mathrm{rad}(K\bar\tau_m)} E(\tau)^{-1+\tau}$

$E(\tau)^{-1+\tau}$

$\mathbf{1}$

$\mathrm{rad}(K\bar\tau_m)$ conjugate

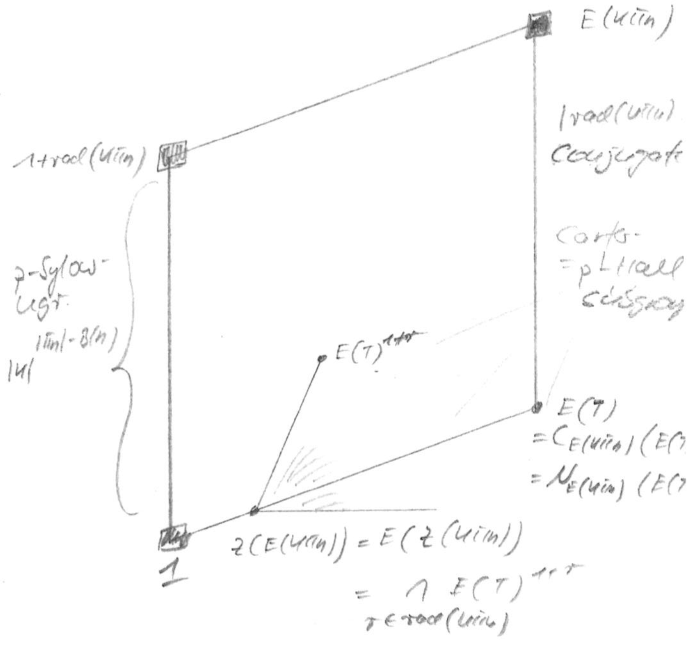

Carta-split conjugacy

$E(K\bar\tau_m)$

$E(\tau)$
$= C_{E(K\bar\tau_m)}(E\tau)$
$= N_{E(K\bar\tau_m)}(E\tau)$

— Cartan subalgebras of $K\bar\tau_m^{\,0}$ —

— p-Sylow Cartan - self-normalizing subalgebra

8.4 Open-ended questions

- Let K be a field and M a finite idempotent monoid. What is the center of KM? Is KM central? Is KM direct-indecomposable?

- Let M be a finite idempotent monoid. What is the center of M?

- Let K be a field and M a finite idempotent monoid. What is the center of $E(KM)$?

- Let K be a field and M a finite idempotent monoid. Describe the centralizer and its dimension of the basis elements M in KM! Clarify the special case $M = \Pi_n$ for an arbitrary element $n \in \mathbb{N}$, too. What is the answer for the basis elements $\{e_P \mid P \in \Pi_n\}$?

- Let K be a field and M a finite idempotent monoid. What is the center of $1 + rad(KM)$? What is the answer for the special case $M = \Pi_n$ for an arbitrary element $n \in \mathbb{N}$?

- Let K be a field and M a finite idempotent monoid. What are the conjugacy classes within $E(KM)$? What are the normal subgroups of $1 + rad(KM)$? What is the answer for the special case $M = \Pi_n$?

8.5 Exercises

Exercise 104 *Let K be a field. Is $K\Pi_3$ central?*

Exercise 105 *Prove (if needed by a research in the literature) the following theorem of Adrian Albert: Let A be a central-simple associative finite-dimensional algebra possessing a non-trivial anti-automorphism. Then A possesses an anti-automorphism of order 2.*

Exercise 106 *Let K be a field. What is the center of $K\Pi_4$?*

Exercise 107 *Let K be a field. Describe the center of $E(K\Pi_5)$!*

Exercise 108 *What is the center of Π_9?*

Exercise 109 *Let K be a field. Is it possible to decompose $K\Pi_9$ as direct product with at least 2 factors?*

Exercise 110 *What is $(13, 564, 2) \vee (789, 10, 11)$?*

Exercise 111 *Let K be a field. What is the intersection of all maximal nilpotent subalgebras of $(K\Pi_9)^\circ$?*

Exercise 112 *Let K be a field. What is the intersection of all Carter subgroups of $E(K\Pi_9)$?*

Exercise 113 *Let K be a field and G a group. How can we use the function $g \mapsto g^{-1}$ of G to define an anti-automorphism on KG?*

Exercise 114 *Let K be a field and $n \in \mathbb{N}$. The transpose-function $A \mapsto A^t$ is an anti-automorphism of $K^{n \times n}$!*

Exercise 115 *How many and which elements are associated to $(132, 45)$ in Π_5?*

Exercise 116 *Let K be a field. Which elements commute with $e_{(12,3)}$ in $K\Pi_3$? Prove that the set of these elements is a subspace and determine its dimension.*

Exercise 117 *Let K be a field of order 3. How many and which elements are commuting with $1 + (12, 3) - (3, 12)$ in $E(K\Pi_3)$?*

Exercise 118 *Let K be a field possessing exactly two elements and $n \leq 3$. What is the center of $1 + rad(K\Pi_n)$? Is the answer identical for an arbitrary field? What is the answer for an arbitrary n?*

Exercise 119 *Let K be a field, $n \in \mathbb{N}$ and A the algebra of lower triangular matrices of $K^{n \times n}$.*

- *What is the center of A?*

- *What is the center of $E(A)$?*

- *Apply the results and analysis of this chapter to A and $E(A)$!*

Chapter 9

Stagnation of central-chains

Let A be a finite-dimensional associative unitary solvable K-algebra possessing a self-centralizing radical complement. This chapter covers the following topics with respect to the stagnation of central chains of $A°$ and $E(A)$:

- Stagnation of the ascending central chain of $A°$ at the center of A

- Stagnation of the descending central chain of $A°$ at the radical of $A/rad(A)$ if A splits over K

- Stagnation of the ascending central chain of $E(A)$ at the center of $E(A)$

- Consequences of part (i), (ii) and (iii) for $K\Pi_n$

- Stagnation of the descending central chain of $E(K\Pi_n)$ and $E(D_n)$ at the derived subgroup by using commutator calculations with Pierce components and a sum-product-lemma that represents a vector-sum in $K\Pi_n$ resp. D_n as a product of units in $E(K\Pi_n)$ resp. $E(D_n)$.

9.1 Stagnation of Lie central chains

Definitions 10 Let L a Lie algebra and G a group. If $S \in \{L, G\}$, then for all $n \in \mathbb{N}_0$ the set $Z_n(S)$ resp. $S^{(n)}$ is the n-th member of the ascending resp. descending central chain of S. In particular, $S' := [S, S] := S^{(1)}$ is the derivation of S. For subsets X, Y of S we denote by $[X, Y]$ the commutator of X and Y. If S is nilpotent, then let $cl(S)$ be its class of nilpotency.⋄

A self-centralizing radical complement contains the center of the algebra. For such algebras the following result is valid:

Theorem 14 *Let K be a field, A a finite-dimensional associative solvable K-algebra and T a radical complement of A containing the center of A. For all $n \in \mathbb{N}$ the identity $Z_n(A^\circ) = Z(A)$ is true.*

Proof: It is sufficient to prove $Z_2(A^\circ) = Z(A)$. By definition the identity $Z(A) \subseteq Z_2(A^\circ)$ is valid. Let $z \in Z_2(A^\circ)$. Thus, $z \circ A \subseteq Z(A) \subseteq T$ is true. We use the solvability of A to derive $z \circ A \subseteq A \circ A \subseteq rad(A)$. Hence, $z \circ A = 0$ and $z \in Z(A)$ are valid.$\diamond$

Corollary 7 *Let K be a field and $n \in \mathbb{N}$. For all $r \in \mathbb{N}$ the identity $Z_r((K\Pi_n)^\circ) = \langle 1 \rangle_K$ is valid.*

Proof: The corollary is a direct consequence of theorems 14, 9 and 13.$\diamond$

Theorem 15 *Let K be a field and A a finite-dimensional associative solvable unitary K-algebra splitting over K for which a self-centralizing radical complement exists. For all $n \in \mathbb{N}$ the identity $(A^\circ)^{(n)} = rad(A)$ is valid.*

Proof: Let $e_1, \cdots, e_n$ be pairwise orthogonal idempotents of A such that $T := \langle e_1, \cdots, e_n \rangle_K$ is a radical complement. If one self-centralizing radical complement exists, then all are self-centralizing based on the theorem of Wedderburn-Malcev. In particular, T is self-centralizing, and by using lemma 5 we derive $rad(A) = \bigoplus_{i \neq j = 1}^{n} e_i A e_j$ and $T = \bigoplus_{i=1}^{n} e_i A e_i$. A is solvable, and thus $A \circ A \subseteq rad(A)$ is true. Let $i, j \in \mathbb{N}$ such that $i \neq j$ and $a \in A$ are valid. We calculate $(e_i a e_j) \circ e_j = e_i a e_j$. Hence, $rad(A) \subseteq rad(A) \circ A \subseteq A \circ A \subseteq rad(A)$ is valid, and we deduct $A \circ A = rad(A) = rad(A) \circ A$.$\diamond$

Corollary 8 *Let K be a field and $n \in \mathbb{N}$. For all $r \in \mathbb{N}$ the identity $((K\Pi_n)^\circ)^{(r)} = rad(K\Pi_n)$ is valid.*

Proof: This corollary is a direct consequence of theorems 15 and 9.$\diamond$

9.2 Stagnation of the ascending central chain of the group of units

Theorem 16 *Let K be a field of order at least 3, A an associative finite-dimensional unitary solvable K-algebra and T a self-centralizing radical complement. For all $n \in \mathbb{N}$ the identity $Z_n(E(A)) = Z(E(A))$ is valid.*

Proof: The algebra A is a semi-direct sum of $rad(A)$ and $T = C_A(T)$. Based on corollary 1.1.8 in [23] we derive that $E(A)$ is a semi-direct product of $1 + rad(A)$ and $E(T) = E(C_A(T))$. A is solvable and T is commutative, and the derivation of $E(A)$ is contained in the normal subgroup $1 + rad(A)$. We only have to prove $Z_2(E(A)) = Z(E(A))$. Let $g \in Z_2(E(A))$. Then $[g, E(A)] \subseteq Z(E(A))$ is valid. K contains at least 3 elements, and by using proposition 13 we derive the statement $Z(E(A)) = E(Z(A))$. Thus, $[g, E(A)] \subseteq E(Z(A)) \subseteq E(C_A(T)) = E(T)$ is valid. In addition, $[g, E(A)] \subseteq [E(A), E(A)] \subseteq 1 + rad(A)$ is true. We conclude $[g, E(A)] \subseteq (1 + rad(A)) \cap E(T) = 1$, and thus g is central in $E(A)$.⋄

Corollary 9 *Let K be a field possessing at least 3 elements and $n \in \mathbb{N}$. For all $r \in \mathbb{N}$ the identity $Z_r(E(K\Pi_n)) = Z(E(K\Pi_n)) \cong K \setminus \{0\}$ is valid.*

Proof: The corollary is a direct consequence of theorems 16 and 9.⋄

Remark 17 Let $K = GF(2)$ and $n \in \mathbb{N}$. The group of units of $K\Pi_n$ consists of the units $1 + rad(K\Pi_n)$ only and is nilpotent. In this case a complete different behaviour is valid as described in corollary 9: the ascending central chain does not stagnate but reaches the whole group. The calculation of the class of nilpotency is done within chapter **Classes of nilpotency and solvability** based on the ascending central chain.⋄

Corollary 10 *Let $n \in \mathbb{N}$ and K a field of characteristic 0. For all $r \in \mathbb{N}$ the identity $Z_r(E(D_n)) = Z(E(D_n))$ is valid.*

Proof: This corollary is a direct consequence of theorems 16 and 5.1 in [3].⋄

9.3 A sum-product-lemma

Definition 1 If A is a K-algebra, then we define $a * b := a + b + ab$ for all $a, b \in A$ and call – as already done by Bartel Leendert van der Waerden – $*$ the star or circle composition on A.⋄[1]

Remark 18 For every associative K-algebra A the following statements are valid:

---

[1] Bartel Leendert van der Waerden

(i) $(A; *)$ is a monoid with zero element 0_A.

(ii) If A is unitary, then the function $A \to A$, $a \mapsto 1_A + a$ is a monoid-isomorphism between $(A; *)$ and $(A; \cdot)$.◇

Definitions 11 If A is an associative K-algebra, then we use $Q(A)$ to symbolize the group of units of the monoid $(A; *)$. For all $a \in Q(A)$ let a' be the inverse of a in $Q(A)$. The elements of $Q(A)$ are called star- or quasi-regular and the group $Q(A)$ is called the star or quasi-regular group of A. If A is unitary, then let $E(A)$ be the group of units of A.◇

Remark 19 For every associative unitary K-algebra A the following statements are valid:

(i) The restrictions of the function $A \to A$, $a \mapsto 1_A + a$ on $Q(A)$ is a group isomorphism between $Q(A)$ and $E(A)$.

(ii) If A is a field, then $E(A) = A \setminus \{0\}$ and $Q(A) = A \setminus \{-1\}$ are true.

(iii) If $a, b \in E(A)$, then $[a, b] = 1 + a^{-1}b^{-1}(a \circ b)$ is valid.◇

Proposition 14 *Let A be an associative unitary K-algebra, $k \in Q(K)$, $e, x, f \in A$, $a := 1 + exf$ and $b := 1 + ke$. The following statements are valid:*

(i) *If e, f are orthogonal, then a is invertible and $a^{-1} = 1 - exf$ is true.*

(ii) *If e is an idempotent of A, then b is invertible and $b^{-1} = 1 + k'e$ is valid.*

(iii) *If e, f are orthogonal and idempotents of A, then $[a, b] = 1 + e((-1)(kk')x)f$ is true.*

Proof: ad(i): This is a direct consequence of the orthogonality of e and f and the 3rd binomial rule: $(1 + exf)(1 - exf) = 1^2 - (exf)^2 = 1 - exfexf = 1 - 0 = 1$.

ad(ii): Because of $k \in Q(K)$ and $e^2 = e$ we calculate:

$$
\begin{aligned}
& (1 + ke)(1 + k'e) \\
= {}& 1 + k'e + ke + kk'e^2 \\
= {}& 1 + (k + k' + kk')e \\
= {}& 1 + (k * k')e \\
= {}& 1 + 0e \\
= {}& 1.
\end{aligned}
$$

ad(iii): We use the idempotency and orthogonality to calculate:

$$a \circ b$$
$$= (1 + exf) \circ (1 + ke)$$
$$= (exf) \circ (ke)$$
$$= (exf) \cdot (ke) - (ke) \cdot (exf)$$
$$= -kexf.$$

Thus, by using (i), (ii) and remark 19 we derive:

$$[a, b]$$
$$= 1 + a^{-1}b^{-1}(a \circ b)$$
$$= 1 + (1 - exf)(1 + k'e)(-kexf)$$
$$= 1 + (1 - exf)(-kexf - k'e^2xf)$$
$$= 1 + (1 - exf)(-1)(k + k')exf$$
$$= 1 - (kk')exf$$
$$= 1 + e((-1)kk'x)f. \diamond$$

Within the following lemma a connection between a vector-sum and a product of these vectors is established. It is the basis for studying the stagnation of the descending central chain of the group of units of D_n and $K\Pi_n$.

Lemma 7 *(sum-product-lemma) Let A be an associative unitary K-algebra, I a finite set, $e_i, i \in I$ pairwise orthogonal idempotents of A, $T := \bigoplus_{i \neq j \in I} e_i A e_j$ and $n := | I |$. A function $f : I \to \mathbb{N}_0$ may exists such that for all $i \neq j \in I$ facing $e_i A e_j \neq 0$ the inequality $f(i) < f(j)$ is valid. For all $i \neq j \in I$ let $a_{i,j} \in A$ and $t := \sum_{i \neq j \in I} e_i a_{i,j} e_j$. Then for all $s \in \underline{n^2}$ an element $x_s \in \{e_i a_{i,j} e_j \mid i \neq j \in I\} \cup \{0\}$ exists such that $1 + t = (1 + x_1) \cdots (1 + x_{n^2})$ is valid.*

Proof: *Step 1:* At first, we re-sort the sum which defines the element t. Let $m := max\{f(i) \mid i \in I\}$. We calculate:

$$1 + t$$
$$= 1 + \sum_{i \neq j \in I} e_i a_{i,j} e_j$$
$$= 1 + \sum_{s=m}^{1} \sum_{\substack{j \in I \\ f(j)=s}} \sum_{\substack{i \in I \setminus \{j\} \\ e_i a_{i,j} e_j \neq 0}} e_i a_{i,j} e_j.$$

Step 2: Let $j \in I$. The idempotents $e_i, i \in I$ are pairwise orthogonal, and thus the following calculation is valid:

$$1 + \sum_{\substack{i \in I \setminus \{j\} \\ e_i a_{i,j} e_j \neq 0}} e_i a_{i,j} e_j$$

$$= \prod_{\substack{i \in I \setminus \{j\} \\ e_i a_{i,j} e_j \neq 0}} (1 + e_i a_{i,j} e_j).$$

Step 3: Let $s \in \underline{m}$, and let $j_{r_1}, \cdots, j_{r_s}$ the elements of I for which the image under f is exactly s. The following equality is true:

$$1 + \sum_{\substack{j \in I \\ f(j) = s}} \sum_{\substack{i \in I \setminus \{j\} \\ e_i a_{i,j} e_j \neq 0}} e_i a_{i,j} e_j$$

$$= \prod_{\substack{i \in I \\ e_i a_{i,j_{r_1}} e_{j_{r_1}} \neq 0}} (1 + e_i a_{i,j_{r_1}} e_{j_{r_1}}) \quad \cdots \quad \prod_{\substack{i \in I \\ e_i a_{i,j_{r_1}} e_{j_{r_s}} \neq 0}} (1 + e_i a_{i,j_{r_s}} e_{j_{r_s}})$$

, because: Let $x, y \in \underline{s}$ such that $x < y$ and $i \in I$ and $e_i a_{i,j_{r_y}} e_{j_{r_y}} \neq 0$. Thus, $e_{j_{r_x}}$ and e_i are different because of $f(i) < f(j_{r_y}) = f(j_{r_x}) = s$. Now we use the orthogonality of the idempotents $e_i, i \in I$ and step 2 to finish step 3.

Step 4: For all $s \in \underline{m}$ let $j_{s,1}, \cdots, j_{s,r_s}$ those elements of I such that their f-value is exactly s. For all $i \neq j \in I$ let $z_{i,j} := e_i a_{i,j} e_j$. The following calculation is valid:

$$1 + \sum_{\substack{s=m}}^{1} \sum_{\substack{j \in I \\ f(j) = s}} \sum_{\substack{i \in I \setminus \{j\} \\ z_{i,j} \neq 0}} e_i a_{i,j} e_j$$

$$= \prod_{\substack{i \in I \\ z_{i,j_{m,1}} \neq 0}} (1 + e_i a_{i,j_{m,1}} e_{j_{m,1}}) \cdots \prod_{\substack{i \in I \\ z_{i,j_{m,r_m}} \neq 0}} (1 + e_i a_{i,j_{m,r_m}} e_{j_{m,r_m}})$$

$$\cdot \prod_{\substack{i \in I \\ z_{i,j_{m-1,1}} \neq 0}} (1 + e_i a_{i,j_{m-1,1}} e_{j_{m-1,1}}) \cdots \prod_{\substack{i \in I \\ z_{i,j_{m-1,r_{m-1}}} \neq 0}} (1 + e_i a_{i,j_{m-1,r_{m-1}}} e_{j_{m-1,r_{m-1}}})$$

$$\cdot \quad \cdots \quad \cdot$$

$$\cdot \prod_{\substack{i \in I \\ z_{i,j_{1,1}} \neq 0}} (1 + e_i a_{i,j_{1,1}} e_{j_{1,1}}) \cdots \prod_{\substack{i \in I \\ z_{i,j_{1,r_1}} \neq 0}} (1 + e_i a_{i,j_{1,r_1}} e_{j_{1,r_1}})$$

, because: Let $1 \leq y \leq t \leq m$, $u \in \underline{r_y}$ and $v \in \underline{r_t}$. We focus on $e_{j_{t,v}}$ and e_i such that $i \in I$ and $e_i a_{i,j_{y,u}} e_{j_{y,u}} \neq 0$. Because of $t \leq y$ and our assumption we derive $f(i) < f(j_{y,u}) = y \leq t = f(j_{t,v})$. Hence, e_i and $e_{j_{t,v}}$ are different. The orthogonality and step 3 finish the proof.$\diamond$

9.4 Stagnation of the descending central chain of the group of units of $K\Pi_n$ and D_n

Theorem 17 *Let $n \in \mathbb{N}$, K a field of order at least 3 and T a radical complement of $K\Pi_n$. For all $r \in \mathbb{N}$ the identity $E(K\Pi_n)^{(r)} = 1 + rad(K\Pi_n) = [1 + rad(K\Pi_n), E(T)]$ is valid.*

Proof: Let $A := K\Pi_n$ and $T := \langle e_P \mid P \in \Pi_n^< \rangle_K$. Based on theorem 3 the set T is a commutative radical complement, and by using theorem 6.4 in [18] we derive $rad(A) = \bigoplus\limits_{\substack{P,Q\in\Pi_n^< \\ Q<P}} e_P A e_Q$. Corollary 1.1.8 in [23] is used to deduce that the normal subgroup $1 + rad(A)$ and the abelian subgroup $E(T)$ are decomposing $E(A)$ semidirect. In particular, $E(A)' \leq 1 + rad(A)$ is valid. Proposition 14 and lemma 7 are used to prove in the following lines that $1 + rad(A)$ is contained in $[1 + rad(A), E(T)]$. This argumentation will finish the proof.

At first, we use proposition 14. Let $P, Q \in \Pi_n^<$, $P \neq Q$ and $a \in A$. We choose an element $k \in Q(K)$ such that $k \neq 1, 0$ is true which exists because of $\mid K \mid \geq 3$. Proposition 14 is used to deduce with $x := (-1)(kk')^{-1}a$ the equation $[1 + e_P x e_Q, 1 + k e_P] = 1 + e_P a e_Q$. By using this fact and again proposition 14 we derive

$(\star)$ $1 + e_P A e_Q \leq [1 + rad(A), E(T)]$.

Now we use lemma 7. Let $P \neq Q \in \Pi_n^<$ and $e_P A e_Q \neq 0$. Based on theorem 6.4 in [18] we derive $Q < P$. Let $Q := (Q_1, \cdots Q_r)$ and $P := (P_1, \cdots P_t)$. By definition for all $i \in \underline{r}$ an element $j \in \underline{t}$ exists such that $Q_i \subseteq P_j$ is true. Q and P are set partitions of $\underline{n}$, and thus $l(P) \leq l(Q)$ must be true. If $l(P) = l(Q)$ would be valid, then P and Q would be associated because of $Q \leq P$. But this fact and $P, Q \in \Pi_n^<$ would cause $P = Q$. Hence, $l(P) < l(Q)$ must be valid. A function f within lemma 7 to be used is the length-function l. By this fact and by using $(\star)$ the proof is finished.$\diamond$

Theorem 18 *Let $n \in \mathbb{N}$, K a field of characteristic zero and T a radical complement of D_n. For all $r \in \mathbb{N}$ the identity $E(D_n)^{(r)} = 1 + rad(D_n) = [1 + rad(D_n), E(T)]$ is valid.*

Proof: We use some results of the thesis [3] of Thorsten Bauer. Let $A := D_n$ be Solomons algebra and $T := H_n = \langle \nu^p \mid p \vdash n \rangle_K$ a radical complement as defined within lemma 3.4 in [3]. Based on lemma 3 the set T is a commutative radical complement, and based on theorem 3.5 in [3] we derive $rad(A) = \bigoplus\limits_{\substack{p,q\vdash n \\ p\neq q}} \nu^p A \nu^q$. The expression $\nu^p A \nu^q$ is different from zero if and only if q is associated to a power-free decomposition of p (see theorem 3.5 in [3]). We use corollary 1.1.8 in [23] to deduce that the normal subgroup $1+rad(A)$ and the abelian subgroup $E(T)$ are decomposing $E(A)$ semidirect. In particular, $E(A)' \leq 1+rad(A)$ is valid. We want to use proposition 14 and lemma 7 to prove that $1+rad(A)$ is contained in $[1+rad(A), E(T)]$ which finishes the proof.

At first, proposition 14 is used. Let $p, q \vdash n$, q a decomposition of p and $a \in A$. We use an element $k \in Q(K)$ such that $k \neq 1, 0$ which is existing because of $\mid K \mid \geq 3$. Based on proposition 14 and $x := (-1)(kk')^{-1}a$ the equation $[1+\nu^p x \nu^q, 1+k\nu^q] = 1+\nu^p a \nu^q$ is valid. Hence, and again by using proposition 14 we derive

$(\star)$ $1 + \nu^p A \nu^q \leq [1 + rad(A), E(T)]$.

Now, lemma 7 is used. Let $p \neq q \vdash n$ and q be associated to a power-free decomposition of p. By definition $\mid p \mid \leq \mid q \mid$ is true. p and q are different partitions and every association-class contains only one partition the length of p must be smaller than the one of q. The function f in lemma 7 is to be used with the length-function $\mid . \mid$. We use this fact and $(\star)$ to finish the proof.$\diamond$

— Cognition of central claims —

$rod(1)$
$= A \circ A$
$= (A^0)^{(n)}$
descending
central claim
of A^0

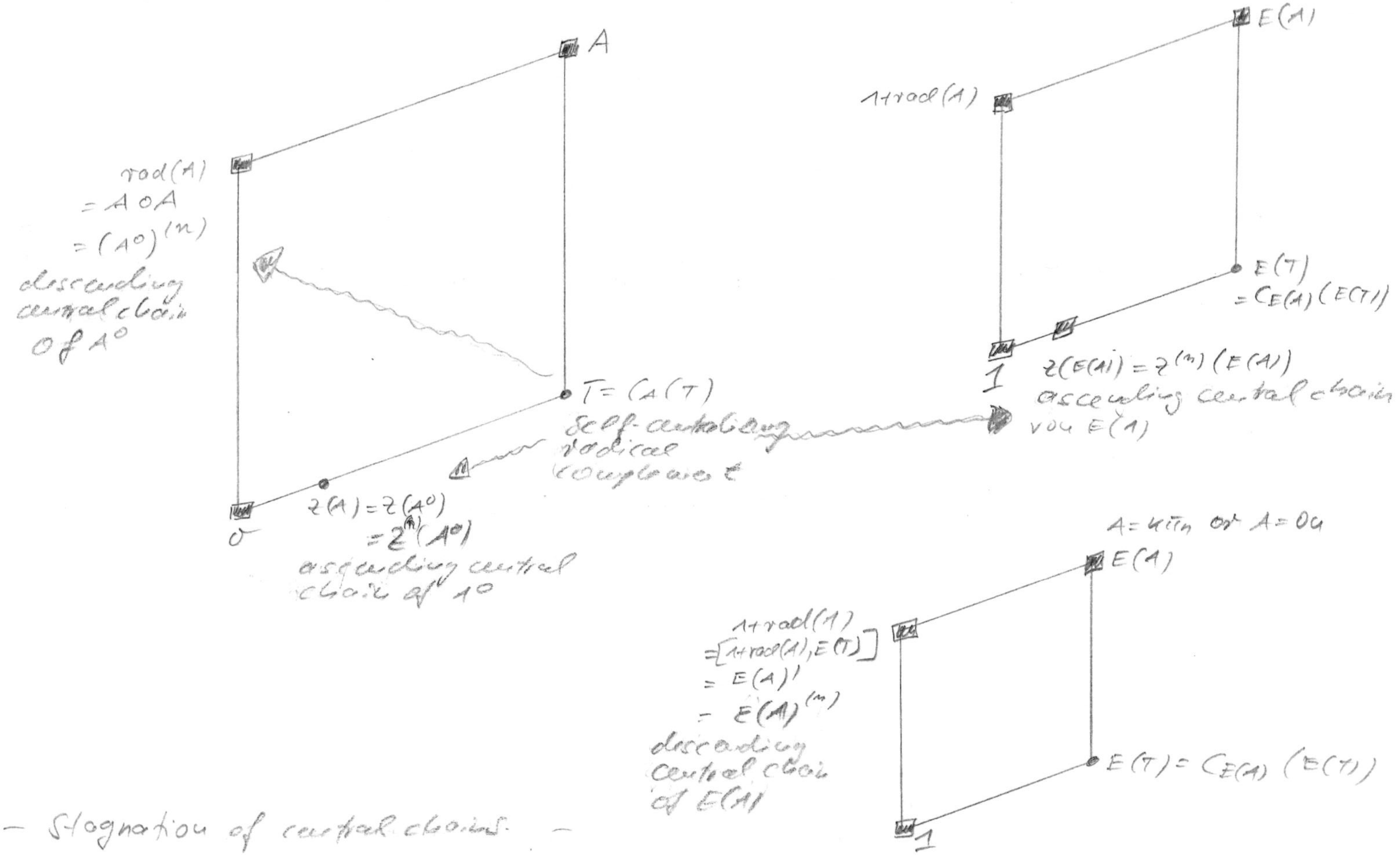

$\zeta(A) = \zeta(A^0)$
$= \zeta(A^0)$
descending central
claim apt 10

$T = (A)(T)$
SE of central claim
radical
compass ζ

$A \to rod(1)$
$= [A \to rod(A), E(1)]$
$= E(A)$
$= E(A)^{(n)}$
descending
central claim
of $E(1)$

$A \to rod(1)$

$\zeta(E(A)) = \zeta^{(n)}(E(A))$
descending central claim
von $E(1)$

$E(T)$
$= E(A)(E(T))$

$E(A)$

$A = 4\tilde{r}m$ on $A = 0a$

$E(T) = E(A)(E(T))$

$E(A)$

9.5 Open-ended questions

- Let K be a field and A a finite-dimensional associative unitary (solvable) K-algebra. Determine the ascending central chain of A° and its length.

- like 1 for the descending central chain

- like 1 but for the group of units

- like 3 for the descending central chain

- True or false: If one of these 4 central chains stagnates after the r-th step, then all 4 chains behaves like this. Compare this question with the theorem of Xiankun Du (see [9]).

- In what way can we transfer the results of this chapter to Solomons algebra in positive characteristic?

9.6 Exercises

Exercise 120 *Prove remark 18!*

Exercise 121 *Prove remark 19!*

Exercise 122 *Let K be a field. Determine the ascending central chain of $(K\Pi_3)^\circ$!*

Exercise 123 *Let K be a field. Determine the descending central chain of $(K\Pi_4)^\circ$!*

Exercise 124 *Let K be a field. Determine the ascending central chain of $E(K\Pi_3)$!*

Exercise 125 *Let K be a field of order ≥ 3. Determine the descending central chain of $E(K\Pi_5)$!*

Exercise 126 *Let K be a field and $k \in K$ such that $k \neq -1$ is valid. Prove that k is quasi-regular and determine its inverse k'. Does a connection exist between k^{-1} and k'? Are $0, 1, -1$ invertible or quasi-regular?*

Exercise 127 *Let A be an associative unitary K-algebra and r a nilpotent element of A. Prove that r is quasi-regular and $1 + r$ is invertible and determine their inverse. Does a connection between these elements exist? (Tip: geometric series and nilpotency)*

Exercise 128 *Let A be an associative K-algebra and I a nilpotent ideal of A. Prove that I is a nilpotent normal subgroup of the star group of A.*

Exercise 129 *Let K be a field of characteristic p and G a finite p-group. Prove $rad(KG) = Aug(KG)$ and the nilpotency of $(KG)^\circ$. In the case $p = 2$ determine the order of KG, $E(KG)$ and $1 + rad(KG)$ for a finite field K and $G = Q_8$ the quaternion group of order 8. What is the class of nilpotency for $(KG)^\circ$ and for $E(KG)$ in this case? (Tip: theorem of Xiankun Du (see [9] and [23]))*

Exercise 130 *Let K be a field and $n \in \underline{3}$. Determine the class of nilpotency of $1 + rad(K\Pi_n)$. Do you have a conjecture for an arbitrary n?*

Exercise 131 *Analyze all topics of this chapter for the algebra of lower triangular matrices over a field! A corresponding sum-product-lemma is contained in [22].*

Chapter 10

Classes of nilpotency and solvability

Let K be a field and A a finite-dimensional associative unitary solvable K-algebra splitting over K such that a self-centralizing radical complement exists. Within this chapter we focus on the following topics with respect to classes of nilpotency and solvability:

- determining the Lie-product of k- and l-fold associative nilradical powers

- determining the descending central chain of $rad(A)^\circ$

- calculating the class of nilpotency of $rad(A)^\circ$ by using the one of $rad(A)$

- determining the chain of derivations of $rad(A)^\circ$, A°, $rad(A)$ and A

- calculating the class of solvability of the previous mentioned structures by using the class of nilpotency of $rad(A)$

- deriving consequences for $K\Pi_n$ and D_n by using theorems of Manfred Schocker and M. D. Atkinson about the associative class of nilpotency of the nilradical

- calculating the commutator of k- and l-fold associative powers of $1 + rad(A)$ for $A = D_n$ and $A = K\Pi_n$ by using the sum-product-lemma

- determining the descending central chain of $1+rad(A)$ for $A = D_n$ and $A = K\Pi_n$

- determining the class of nilpotency for $1 + rad(A)$ for $A = D_n$ and $A = K\Pi_n$ by using the class of nilpotency for $rad(A)$

113

- determining the chain of derivations for the group of units of $A = D_n$ and $A = K\Pi_n$ and for $1 + rad(A)$

- determining the class of solvability for the previous mentioned structures

- determining exponents alongside some chains within $1 + rad(K\Pi_n)$.

10.1 Classes of nilpotency and solvability of the associated Lie algebra

Definitions 12 Let S be a group or a K-Lie algebra. For all $n \in \mathbb{N}_0$ let $S^{[n]}$ be the n-th member of the (descending) chain of derivation of S. If A is an associative K-algebra, then this chain is defined as follows: $A^{[0]} := A$, $A^{[1]} := A'$ (the ideal generated by $A \circ A$ in A) and $A^{[n]} := (A^{[n-1]})'$ for all $n \in \mathbb{N}_{\geq 2}$.

If S or A is solvable, then we denote by $st(S)$ resp. $st(A)$ the class of solvability of S resp. of A. The descending chain of derivations is the fastest chain with abelian factors. Associative powers of A of length $k \in \mathbb{N}$ are denoted by $A^{<k>}$. This set is spanned by all k-fold associative products of k elements.◇

Theorem 19 *Let K be a field and A a finite-dimensional associative unitary solvable K-algebra splitting over K such that a self-centralizing radical complement exists. For all $k, l \in \mathbb{N}$ the statement*

$$rad(A)^{<k>} \circ rad(A)^{<l>} = rad(A)^{<k+l>}$$

is valid.

Proof: Let $k, l \in \mathbb{N}$. The statement $rad(A)^{<k>} \circ rad(A)^{<l>} \subseteq rad(A)^{<k+l>}$ is valid. Let T be a self-centralizing radical complement, like $T = \langle e_1, \cdots, e_n \rangle_K$ spanned by pairwise orthogonal idempotents. Based on lemma 5 we derive $rad(A) = \bigoplus_{i \neq j = 1}^{n} e_i A e_j$ and $T = \bigoplus_{i=1}^{n} e_i A e_i$. For all $i \neq j \in \underline{n}$ let $B_{i,j}$ a K-basis of the Pierce component $e_i A e_j$. The $k + l$-th power of the nilradical is K-spanned by $k + l$-fold associative products, and its factors are elements of the basis $B_{i,j}$ such that $i \neq j \in \underline{n}$. We only need to prove that all these generators are contained in $rad(A)^{<k>} \circ rad(A)^{<l>}$. Let $x := b_{i_1,j_1} \cdots b_{i_k,j_k}$ and $y := b_{i_{k+1},j_{k+1}} \cdots b_{i_{k+l},j_{k+l}}$ such that $b_{i_s,j_s} \in B_{i_s,j_s}$ for all $s \in \underline{k+l}$. We have to prove $xy \in rad(A)^{<k>} \circ rad(A)^{<l>}$. For $j_{k+l} = i_1$ the statement

$xy \in rad(A) \cap (e_{i_1} A e_{i_1}) \leq rad(A) \cap T = \{0\} \leq rad(A)^{<k>} \circ rad(A)^{<l>}$ is valid. Let $j_{k+l} \neq i_1$. Then, $yx = 0$ is true, and we have proven $xy = x \circ y \in rad(A)^{<k>} \circ rad(A)^{<l>}$. $\diamond$

Corollary 11 *Let K be a field and A a finite-dimensional associative unitary solvable K-algebra splitting over K such that a self-centralizing radical complement exists. The following statements are valid:*

(i) *For all $n \in \mathbb{N}$ the identity $rad(A)^{<n>} = (rad(A)^\circ)^{(n)}$ is true.*

(ii) $cl(rad(A)) = cl(rad(A)^\circ) = cl(rad(A)^\star)$

(iii) *For all $n \in \mathbb{N}$ the identity $(rad(A)^\circ)^{[n]} = rad(A)^{<2^n>}$ is valid.*

(iv) $st(rad(A)) = st(rad(A)^\circ) = min\{l \in \mathbb{N} \mid 2^l \geq cl(rad(A))\} = \lfloor log_2(cl(rad(A))$

(v) *For all $n \in \mathbb{N}$ the identity $(A^\circ)^{[n]} = rad(A)^{<2^{n-1}>}$ is true.*

(vi) $st(A) = st(A^\circ) = 1 + st(rad(A)) = 1 + min\{l \in \mathbb{N} \mid 2^l \geq cl(rad(A))\} = 1 + \lfloor log_2(cl(rad(A))) \rfloor$

Proof: ad(i): This part is a direct consequence of theorem 19.
ad(ii): The first equality is derivable by part (i), the second one is a consequence of the theorem of Xiankun Du (see [9]).
ad(iii): This part is a consequence of theorem 19, too.
ad(iv): Part (iii) implies part (iv).
ad(v): Part (v) is deductable by part(iii) and theorem 15 which implies $A \circ A = rad(A)$.
ad(vi): Part (vi) is a consequence of parts (v) and (iv). $\diamond$

Corollary 12 *Let K be a field and $n \in \mathbb{N}$. The following statements are valid:*

(i) *For all $r \in \mathbb{N}$ the identity $rad(K\Pi_n)^{<r>} = (rad(K\Pi_n)^\circ)^{(r)}$ is valid.*

(ii) $cl(rad(K\Pi_n)) = cl(rad(K\Pi_n)^\circ) = cl(rad(K\Pi_n)^\star) = n$

(iii) *For all $r \in \mathbb{N}$ the identity $(rad(K\Pi_n)^\circ)^{[r]} = rad(K\Pi_n)^{<2^r>}$ is true.*

(iv) $st(rad(K\Pi_n)) = st(rad(K\Pi_n)^\circ) = min\{l \in \mathbb{N} \mid 2^l \geq n\} = \lfloor log_2(n) \rfloor$

(v) *For all $r \in \mathbb{N}$ the identity $(K\Pi_n^\circ)^{[r]} = rad(K\Pi_n)^{<2^{r-1}>}$ is valid.*

(vi) $st(K\Pi_n) = st(K\Pi_n^\circ) = 1 + st(rad(K\Pi_n)) = 1 + min\{l \in \mathbb{N} \mid 2^l \geq n\} = 1 + \lfloor log_2(n) \rfloor$.

n	$cl(rad(K\Pi_n))$	$cl(rad(K\Pi_n)^\circ)$	$st(rad(K\Pi_n))$	$st(rad(K\Pi_n)^\circ)$	$st(K\Pi_n)$	$st(K\Pi_n^\circ)$
1	1	1	1	1	2	2
2	2	2	1	1	2	2
3	3	3	2	2	3	3
4	4	4	2	2	3	3
5	5	5	3	3	4	4
6	6	6	3	3	4	4
7	7	7	3	3	4	4
8	8	8	3	3	4	4
9	9	9	4	4	5	5
10	10	10	4	4	5	5
11	11	11	4	4	5	5
12	12	12	4	4	5	5
13	13	13	4	4	5	5
14	14	14	4	4	5	5
15	15	15	4	4	5	5
16	16	16	4	4	5	5

Table 10.1: classes of nilpotency and solvability for $K\Pi_n$ and $(K\Pi_n)^\circ$

Proof: Based on lemma 5 and lemma 6 the radical complement $\langle e_P \mid P \in \Pi_n^\leq \rangle_K$ is self-centralizing. Manfred Schocker has proven within corollary 7.4 in [18] that the class of nilpotency of $rad(K\Pi_n)$ is exactly n. Corollary 11 finishes the proof.$\diamond$

Within table 10.1 we list some classes of nilpotency and solvability for the Solomon-Tits algebra.

Corollary 13 *Let K be a field of characteristic zero and $n \in \mathbb{N}$. The following statements are valid:*

(i) *For all $r \in \mathbb{N}$ the identity $rad(D_n)^{<r>} = (rad(D_n)^\circ)^{(r)}$ is true.*

(ii) *$cl(rad(D_n)) = cl(rad(D_n)^\circ) = cl(rad(D_n)^\star) = n - 1$*

(iii) *For all $r \in \mathbb{N}$ the identity $(rad(D_n)^\circ)^{[r]} = rad(D_n)^{<2^r>}$ is valid.*

(iv) *$st(rad(D_n)) = st(rad(D_n)^\circ) = min\{l \in \mathbb{N} \mid 2^l \geq n-1\} = \lfloor log_2(n-1) \rfloor$.*

(v) *For all $r \in \mathbb{N}$ the identity $(D_n^\circ)^{[r]} = rad(D_n)^{<2^{r-1}>}$ is true.*

(vi) *$st(D_n) = st(D_n^\circ) = 1 + st(rad(D_n)) = 1 + min\{l \in \mathbb{N} \mid 2^l \geq n - 1\} = \lfloor log_2(n-1) \rfloor$.*

n	$cl(rad(D_n))$	$cl(rad(D_n)^\circ)$	$st(rad(D_n))$	$st(rad(D_n)^\circ)$	$st(D_n)$	$st(D_n^\circ)$
1	0	0	0	0	1	1
2	1	1	1	1	2	2
3	2	2	1	1	2	2
4	3	3	2	2	3	3
5	4	4	2	2	3	3
6	5	5	3	3	4	4
7	6	6	3	3	4	4
8	7	7	3	3	4	4
9	8	8	3	3	4	4
10	9	9	4	4	5	5
11	10	10	4	4	5	5
12	11	11	4	4	5	5
13	12	12	4	4	5	5
14	13	13	4	4	5	5
15	14	14	4	4	5	5
16	15	15	4	4	5	5
17	16	16	4	4	5	5

Table 10.2: classes of nilpotency and solvability for D_n and $(D_n)^\circ$

Proof: Based on lemma 5 and part (f) of theorem 3.5 in [3] a self-centralizing radical complement exists. M. D. Atkinson has proven within [2] that the nilradical of D_n is of nilpotency class $n-1$. Now we use corollary 11 to finish the proof.$\diamond$

Within table 10.2 we list some classes of nilpotency and solvability for Solomons algebra.

10.2 Classes of nilpotency and solvability for the group of units of $K\Pi_n$ and D_n

Within this section we focus on the class of solvability of the group of units and the normal subgroup $1 + rad(K\Pi_n)$ resp. $1 + rad(D_n)$. In addition, we determine the descending central chain of these normal subgroups. As a consequence, we calculate the class of nilpotency for them.

Proposition 15 *Let A be an associative unitary K-algebra and r, s nilpotent elements of class 2 of A such that $rsr - srs = 0$ is true. $1 + r, 1 + s$ are units*

118

of A, rs is nilpotent of class 2 and $[1 + r, 1 + s] = 1 + r \circ s$ is valid.

Proof: r, s are nilpotent, and thus $1 + r, 1 + s$ are units of A. Because of $r^2 = 0 = s^2$ we calculate $(1 + r)^{-1} = 1 - r$ and $(1 + s)^{-1} = 1 - s$. Thus,

$$
\begin{aligned}
& [1 + r, 1 + s] \\
= \ & (1 - r)(1 - s)(1 + r)(1 + s) \\
= \ & ((1 + rs) - (s + r))(1 + rs) + (s + r))
\end{aligned}
$$

is valid. Let $a := 1 + rs$ and $b := s + r$. We prove $ab = ba$:

$$
\begin{aligned}
& ab \\
= \ & (1 + rs)(s + r) \\
= \ & s + r + rss + rsr \\
= \ & s + r + rsr \\
= \ & s + r + srs
\end{aligned}
$$

and

$$
\begin{aligned}
& ba \\
= \ & (s + r)(1 + rs) \\
= \ & s + r + srs + rrs \\
= \ & s + r + srs.
\end{aligned}
$$

Now we use the 3rd rule of the binomial theorem:

$$
\begin{aligned}
& [1 + r, 1 + s] \\
= \ & a^2 - b^2 \\
= \ & 1 + rs + rs + rsrs - s^2 - sr - rs - r^2 \\
= \ & 1 + rs + rsrs - sr \\
= \ & 1 + r \circ s + rsrs.
\end{aligned}
$$

We only need to prove $rsrs = 0$. For this, we use $rsr - srs = 0$, and thus $rsrs = srss = 0$ is valid because $s^2 = 0$ is assumed.◇

Proposition 16 *Let A be an associative unitary K-algebra and $k, l \in \mathbb{N}$. The sets $1 + rad(A)^{<k>}$ and $1 + rad(A)^{<l>}$ are subgroups of $1 + rad(A)$ such that $[1 + rad(A)^{<k>}, 1 + rad(A)^{<l>}] \leq 1 + rad(A)^{<k+l>}$ is valid.*

Proof: Let $x \in rad(A)^{<k>}$ and x' its star-inverse. Thus, $(1+x)^{-1} = 1 + x'$ and $x' = -x'x - x \in rad(A)^{<k>}$ are valid because $rad(A)^{<k>}$ is an ideal. Therefor, $1 + rad(A)^{<k>}$ is multiplicative closed. $1 \in 1 + rad(A)^{<k>}$ is valid, and thus $1 + rad(A)^{<k>}$ is a subgroup of $1 + rad(A)$.

Let $y \in rad(A)^{<k>}$, and let y' its star-inverse. We derive $(1+y)^{-1} = 1 + y'$, $x, x' \in rad(A)^{<k>}$ and $y, y' \in rad(A)^{<l>}$. We calculate:

$$
\begin{aligned}
& [1+x, 1+y] \\
= \ & (1+x')(1+y')(1+x)(1+y) \\
= \ & (1+x'+y'+x'y')(1+x+y+xy) \\
= \ & 1 + x + y + xy + x' + x'x + x'y + x'xy + y' + y'x + y'y \\
+ \ & y'xy + x'y' + x'y'x + x'y'y + x'y'xy \\
= \ & 1 + xy + x'y + x'xy + y'x + y'xy + x'y' + x'y'x + x'y'y + x'y'xy \\
= \ & 1 + y'x + y'xy + x'y' + x'y'x + x'y'y + x'y'xy \in rad(A)^{<k+l>}.\diamond
\end{aligned}
$$

Theorem 20 *Let K be a field and $n, k, l \in \mathbb{N}$. The identity*

$$[1 + rad(K\Pi_n)^{<k>}, 1 + rad(K\Pi_n)^{<l>}] = 1 + rad(K\Pi_n)^{<k+l>}$$

is valid.

Proof: One inclusion is the content of proposition 16. Based on theorem 19 the identity $1 + rad(K\Pi_n)^{<k+l>} = 1 + (rad(K\Pi_n)^{<k>} \circ rad(K\Pi_n)^{<l>})$ is valid. We use theorem 9 to deduce $rad(K\Pi_n) = \bigoplus_{P,Q \in \Pi_n^<} e_P K\Pi_n e_Q$. For all $P, Q \in \Pi_n^<$ the Pierce-component for $P \neq Q$ is nilpotent of class 2 and is K-linear spanned by nilpotent elements of class 2. If a is an element of $rad(K\Pi_n)^{<k>} \circ rad(K\Pi_n)^{<l>}$, then a finite subset I of $\mathbb{N}$ and nilpotent elements a_i, a_j of class 2 and of form $e_P a e_Q$ exist such that $P \neq Q \in \Pi_n^<$ and $a = \sum_{i,j \in I} a_i \circ a_j$ are valid. We use the sum-product-lemma 7 and the same argumentation is done within theorem 17 to prove

$$
\begin{aligned}
& 1 + a \\
= \ & 1 + \sum_{i,j \in I} a_i \circ a_j \\
= \ & \prod_{i,j \in I} (1 + (a_i \circ a_j)).
\end{aligned}
$$

We only have to prove that for all $i, j \in I$ the equality $1 + a_i \circ a_j = [1 + a_i, 1 + a_j]$ is valid. For this, we use proposition 15. Let $i, j \in I$. We can assume that $a_i = e_P x e_Q$ and $a_j = e_S y e_T$ are valid for suitable $x, y \in K\Pi_n$

and $P, Q, R, S \in \Pi_n^{\leq}$ such that $P \neq Q$ and $R \neq S$ are true. Because of proposition 15 we only need to prove $a_i a_j a_i - a_j a_i a_j = 0$. We will prove that $a_i a_j a_i = a_j a_i a_j = 0$ is valid.

If $Q \neq R$ or $S \neq P$, then we use the orthogonality and the idempotency of $e_Z, Z \in \Pi_n^{\leq}$ to deduce $a_i a_j a_i = a_j a_i a_j = 0$. Let $Q = R$ and $S = P$. Then, $a_i a_j a_i = e_P x e_Q y e_P x e_Q$ and $a_j a_i a_j = e_Q y e_P x e_Q y e_P$ are valid. These products are zero because one of the Pierce components $e_P K \Pi_n e_Q$ and $e_Q K \Pi_n e_P$ is zero and the products contains factors of these two components. If both components are not zero, then we would derive from page 27 in [18] $P \leq Q$ and $Q \leq P$. P, Q are associated, and based on the definition of $\Pi_n^{\leq}$ this would cause $P = Q$ which is contradicting the choice of P, Q.$\diamond$

Corollary 14 *Let K be a field and $n \in \mathbb{N}$. The following statements are valid:*

(i) *The descending central chain of $1 + rad(K \Pi_n)$ is $\{1 + rad(K \Pi_n)^{<l>} \mid l \in \underline{n}\}$.*

(ii) *The chain of derived subgroups of $1 + rad(K \Pi_n)$ is $\{1 + rad(K \Pi_n)^{<2^k>} \mid k \in \mathbb{N}\}$.*

(iii) *The class of solvability of $1 + rad(K \Pi_n)$ is $st(rad(K \Pi_n)) = st(rad(K \Pi_n)^{\circ}) = min\{l \in \mathbb{N} \mid 2^l \geq n\} = \lfloor log_2(n) \rfloor$.*

(iv) *If K possesses at least 3 elements, then the chain of derived subgroups of $E(K \Pi_n)$ is $\{1 + rad(K \Pi_n)^{<2^k>} \mid k \in \mathbb{N}_0\}$.*

(v) *If K possesses at least 3 elements, then the class of solvability of $E(K \Pi_n)$ is identical to one of $K \Pi_n$ and $(K \Pi_n)^{\circ}$: $st(K \Pi_n) = st(K \Pi_n^{\circ}) = 1 + min\{l \in \mathbb{N} \mid 2^l \geq n\} = 1 + \lfloor log_2(n) \rfloor$.*

Proof: The proof is a direct consequence of corollary 12 and of the theorems 20 and 17.$\diamond$

Table 10.3 lists some classes of nilpotency and solvability for the Solomon-Tits algebra.

Theorem 21 *Let K be a field of characteristic zero and $n, k, l \in \mathbb{N}$. The identity*

$$[1 + rad(D_n)^{<k>}, 1 + rad(D_n)^{<l>}] = 1 + rad(D_n)^{<k+l>}$$

is valid.

n	$st(1+rad(K\Pi_n))$	$st(E(K\Pi_n))$	$st(rad(K\Pi_n))$	$st(rad(K\Pi_n)^\circ)$	$st(K\Pi_n)$	$st(K\Pi_n^\circ)$
1	1	2	1	1	2	2
2	1	2	1	1	2	2
3	2	3	2	2	3	3
4	2	3	2	2	3	3
5	3	4	3	3	4	4
6	3	4	3	3	4	4
7	3	4	3	3	4	4
8	3	4	3	3	4	4
9	4	5	4	4	5	5
10	4	5	4	4	5	5
11	4	5	4	4	5	5
12	4	5	4	4	5	5
13	4	5	4	4	5	5
14	4	5	4	4	5	5
15	4	5	4	4	5	5
16	4	5	4	4	5	5

Table 10.3: classes of nilpotency and solvability for $K\Pi_n$

Proof: One inclusion is proven within proposition 16.

We use some results and notation of the thesis [3] of Thorsten Bauer about the structure of Solomons algebra. Let $A := D_n$ be Solomons algebra and $T := H_n = \langle \nu^p \mid p \vdash n \rangle_K$ a radical complement as defined and constructed within lemma 3.4 in [3]. Lemma 3 lets us deduce that T is commutative, and based on theorem 3.5 in [3] we deduct $rad(A) = \bigoplus_{\substack{p,q \vdash n \\ p \neq q}} \nu^p A \nu^q$. Within this

expression the Pierce component $\nu^p A \nu^q$ is zero if and only if q is associated to a power-free decomposition of p associated (see theorem 3.5 in [3]). We use theorem 19 to derive $1 + rad(D_n)^{<k+l>} = 1 + (rad(D_n)^{<k>} \circ rad(D_n)^{<l>})$. For all p, q such that q is associated to a power-free decomposition of p the Pierce component $\nu^p A \nu^q$ is nilpotent of class 2 and is K-linear spanned by nilpotent elements of class 2 linear. If a is an element of $rad(D_n)^{<k>} \circ rad(D_n)^{<l>}$, then a finite subset I of $\mathbb{N}$ and nilpotent elements a_i, a_j of class 2 and form $\nu^p x \nu^q$ exists (q is again associated to a power-free decomposition of p) such that $a = \sum_{i,j \in I} a_i \circ a_j$ is valid. Now we use the sum-product-lemma 7 and a similar argumentation as done within theorem 18 to derive

$$1 + a$$

$$= 1 + \sum_{i,j \in I} a_i \circ a_j$$

$$= \prod_{i,j \in I} (1 + (a_i \circ a_j)).$$

We only have to rove that for all $i, j \in I$ the identity $1 + a_i \circ a_j = [1 + a_i, 1 + a_j]$ is valid. For this, we use proposition 15. Let $i, j \in I$. We can assume that $a_i = \nu^p x \nu^q$ and $a_j = \nu^s y \nu^t$ is valid such that $x, y \in D_n$ and p, q, s, t (q is again associated to a power-free decomposition of p and t to one of s). We only have to prove – based on proposition 15 – that $a_i a_j a_i - a_j a_i a_j = 0$ is valid. Indeed, we will prove $a_i a_j a_i = a_j a_i a_j = 0$.

If $q \neq r$ or $s \neq p$, then we use the orthogonality of the idempotents – which are decomposing 1 as a sum – that $a_i a_j a_i = a_j a_i a_j = 0$ is true. Let $q = r$ and $s = p$ be valid. In this case $a_i a_j a_i = \nu^p x \nu^q y \nu^p x \nu^q$ and $a_j a_i a_j = \nu^q y \nu^p x \nu^q y \nu^p$ are true. This products are zero because at least one of the Pierce components $\nu^p D_n \nu^q$ and $\nu^q D_n \nu^p$ are zero and the factors within the product are contained in both Pierce components. If both Pierce components would not be zero, then theorem 3.5 in [3] would yield to the fact that p is associated to a power-free decomposition of q and vice versa. By definition of a power-free decomposition the length of p and q differ by 1 which is a contradiction.◇

Corollary 15 *Let K be a field, $char(K) = 0$ and $n \in \mathbb{N}$. The following statements are valid:*

(i) The descending central chain of $1 + rad(D_n)$ is $\{1 + rad(D_n)^{<l>} \mid l \in \underline{n}\}$.

(ii) The chain of derived subgroups of $1 + rad(D_n)$ is $\{1 + rad(D_n)^{<2^k>} \mid k \in \mathbb{N}\}$.

(iii) The class of solvability of $1 + rad(D_n)$ is exactly $st(rad(D_n)) = st(rad(D_n)^\circ)$ $min\{l \in \mathbb{N} \mid 2^l \geq n\} = \lfloor log_2(n-1) \rfloor$.

(iv) The chain of derived subgroups of $1 + rad(D_n)$ is $\{1 + rad(D_n)^{<2^k>} \mid k \in \mathbb{N}_0\}$.

(v) The class of solvability of the group of units of D_n is identical to the one of D_n and of $(D_n)^\circ$: $st(D_n) = st(D_n^\circ) = 1 + min\{l \in \mathbb{N} \mid 2^l \geq n\} = 1 + \lfloor log_2(n-1) \rfloor$.

Proof: The proof is a direct consequence of corollary 13 and the theorems 20 and 18.◇

Table 10.4 lists some classes of nilpotency and solvability for the Solomon algebra.

n	$st(1+rad(D_n))$	$st(E(D_n))$	$st(rad(D_n))$	$st(rad(D_n)^\circ)$	$st(D_n)$	$st(D_n^\circ)$
1	0	1	0	0	1	1
2	1	2	1	1	2	2
3	1	2	1	1	2	2
4	2	3	2	2	3	3
5	2	3	2	2	3	3
6	3	4	3	3	4	4
7	3	4	3	3	4	4
8	3	4	3	3	4	4
9	3	4	3	3	4	4
10	4	5	4	4	5	5
11	4	5	4	4	5	5
12	4	5	4	4	5	5
13	4	5	4	4	5	5
14	4	5	4	4	5	5
15	4	5	4	4	5	5
16	4	5	4	4	5	5
17	4	5	4	4	5	5

Table 10.4: classes of nilpotency and solvability for D_n

10.3 Exponents alongside some chains within $1 + rad(K\Pi_n)$

Let K be a field and $n \in \mathbb{N}$. Within corollary 22 we have described the descending central chain and the chain of derived subgroups for the nilpotent group $1+rad(K\Pi_n)$ by powers of the nilradical. This result has an immediate consequence to the factors of these two chains in the case of a finite field of characteristic p: all factors are elementary-p-abelian. The size of the radical powers is known by a theorem of Manfred Schocker. Thus, we can describe the factors completely. The term factor is used for factor groups alongside a chain $(1 = S_1, S_2, ..., S_n = G)$ of a group G of length n such that all S_i are subgroups and S_i is a normal subgroup of S_{i+1} for all $i \in \underline{n-1}$. Factors are the quotient groups S_{i+1}/S_i for all $i \in \underline{n-1}$. After proving this result we examine also the ascending central chain within the nilpotent group $1+rad(K\Pi_n)$ in the case of a finite field of characteristic p and prove – without knowledge of the centers the same result: all factors are elementary-p-abelian. The size or any kind of descriptions of the centers is not known to the author.

Theorem 22 *Let K be a finite field of characteristic p and $n \in \mathbb{N}$. The following statements are valid:*

(i) *All factors alongside the descending central chain of $1 + rad(K\Pi_n)$ are elementary-p-abelian.*

(ii) *All factors alongside the chain of derived subgroups of $1 + rad(K\Pi_n)$ are elementary-p-abelian.*

Proof: Based on corollary 22 the descending central chain of $1 + rad(K\Pi_n)$ is exactly $\{1 + rad(K\Pi_n)^{<l>} \mid l \in \underline{n}\}$ and the chain of derived subgroups of $1 + rad(K\Pi_n)$ is identical to $\{1 + rad(K\Pi_n)^{<2^k>} \mid k \in \mathbb{N}\}$. Thus, we have to investigate factors of type $1 + rad(K\Pi_n)^{<l>}/1 + rad(K\Pi_n)^{<l+1>}$ and $1 + rad(K\Pi_n)^{<2^k>}/1 + rad(K\Pi_n)^{<2^{k+1}>}$ for $k, l \in \mathbb{N}$. Let $k \in \mathbb{N}$ and $a \in 1+rad(K\Pi_n)^{<k>}$. At first, we use the well-known fact that within characteristic p for two commuting elements x, y the identity $(x + y)^p = x^p + y^p$ is valid. Thus, $(1 + a)^p = 1^p + a^p = 1 + a^p$ is true. In addition, we derive $a^p \in (rad(K\Pi_n)^{<k>})^p = rad(K\Pi_n)^{<pk>}$. Because of $pk \geq max\{k + 1, 2k\}$ both parts are proven.$\diamond$

Let $P, Q \in \Pi_n^{\leq}$ and $Q \leq P$. We define $c_{P,Q} := dim(e_P K\Pi_n e_Q)$. Manfred Schocker investigates the associative powers of the nilradical and relates them to these special Pierce components. By using this theorem the size of the factors is determinable which is illustrated within the examples. The following theorem summarizes parts of theorem 6.4 and corollaries 7.4 and 7.5 in [18]:

Theorem 23 *(Manfred Schocker, 2006) Let K be a field, $n \in \mathbb{N}$ and $k \in \underline{n-1}_]$. The following statements are valid:*

(i) *The nilpotency class (also called nilindex) of $rad(K\Pi_n)$ is exactly n.*

(ii) *$rad(K\Pi_n)^{<n-1>} = e_{(123...n)} K\Pi_n e_{(1,2,3,...,n)}$ is of dimension $(n-1)!$.*

(iii) *$rad(K\Pi_n)^{<k>} = rad(K\Pi_n)^{<k+1>} \bigoplus\limits_{Q\in\Pi_n^{<}} \bigoplus\limits_{Q\leq T\in\Pi_n^{<},l(Q)-l(T)=k} e_T K\Pi_n e_Q$*

(iv) *Let $P,Q \in \Pi_n^{<}$, $Q \leq P$, $P := (P_1,\cdots,P_l)$ and $Q := (Q_1,\cdots,Q_k) \in \Pi_n^{<}$. Then $c_{P,Q} = \prod\limits_{j=1}^{l}(m_j-1)!$ with $m_j := | \{i \mid i \in \underline{k}_], Q_i \subseteq P_j\} |$ is valid. In particular $\sum\limits_{j=1}^{l} m_j = k$ is true.$\diamond$*

By using this theorem we can describe the size of the factors alongside the descending central chain and the chain of derivations of $1+rad(K\Pi_n)$. Recall that the rank of an elementary-abelian p-group G for a prime number p is the number of cyclic p-groups C_p of order p in a decomposition of G into C_p-groups.

Corollary 20 *Let p be a prime number, $l \in \mathbb{N}$, K a finite field of characteristic p and order p^l, $n \in \mathbb{N}$ and $k \in \underline{n-1}_]$. The following statements are valid:*

(i) *The rank of the elementary-abelian p-group $1+rad(K\Pi_n)^{<k>}/1+rad(K\Pi_n)^{<k}$ is exactly $l \cdot \sum\limits_{Q<P\in\Pi_n^{<},l(Q)-l(P)=k} c_{P,Q}.$*

(ii) *The rank of the elementary-abelian p-group $1 + rad(K\Pi_n)^{<2^k>}/1 + rad(K\Pi_n)^{<2^{k+1}>}$ is exactly $l \cdot \sum\limits_{Q<P\in\Pi_n^{<},l(Q)-l(P)=2^k,2^k+1,...,2^{k+1}-1} c_{P,Q}.$*

(iii) *Both ranks are growing with n.*

Proof: ad(i)+(ii): Both parts are a direct consequence of theorem 23.

ad(iii): Let $P,Q \in \Pi_n^{<}$, $Q \leq P$, $P := (P_1,\cdots,P_l)$ and $Q := (Q_1,\cdots,Q_k) \in$

$\Pi_n^<$. Then $(P, n+1)$ and $(Q, n+1)$ are elements of $\Pi_{n+1}^<$ for which $(Q, n+1) < (P, n+1)$ is still valid. Based on part (iv) of theorem 23 the identity $c_{P,Q} = c_{(P,n+1),(Q,n+1)}$ is valid. Thus, by using again theorem 23 the dimension of $rad(K\Pi_n)^{<k>}/rad(K\Pi_n)^{<k+1>}$ is not greater than the one of $rad(K\Pi_{n+1})^{<k>}/rad(K\Pi_{n+1})^{<k+1>}$ (also because the map $P \mapsto (P, n+1)$ is injective). Indeed, the dimension is smaller, because $(123....n(n+1))$ is minimal in $\Pi_{n+1}^<$, not identical to $(P, n+1)$ and for each $s \in \underline{n+1}$ an element S in $\Pi_{n+1}^<$ of length s exists (see exercises). By using theorem 23 the dimension of $e_S K\Pi_{n+1} e_{(1,2,3,...,n,n+1)}$ is at least 1. Hence, we can use the parts (i) and (ii) to finish the proof.◇

Example 8 Let K be a field and $n \in \mathbb{N}$. Manfred Schocker has proven within corollary 7.4 in [18] that the class of nilpotency of $rad(K\Pi_n)$ is exactly n. Within example 1 we have determined the dimension of the radical and of an radical complement for $n = 1, 2, 3$. For $n = 1$ the radical of $K\Pi_1$ is zero, for $n = 2$ the dimension of $rad(K\Pi_2)$ is 1 and for $n = 3$ this dimension is 8. Let $J_3 := rad(K\Pi_3)$. We use part (ii) of theorem 23 to deduce that the dimension of J_3^2/J_3^3 is exactly $(3 - 1! = 2$. Thus, the dimension of J_3/J_3^2 is exactly $8 - 2 = 6$.

Let $J_4 = rad(K\Pi_4)$. Within chapter 2 and 3 we have calculated the dimension of $K\Pi_4$ and its radical factor structure: 70 and 15. Thus, J_4 is of dimension $70 - 15 = 55$. We use part (ii) of theorem 23 to deduce that the dimension of J_4^3/J_4^4 is exactly $(4 - 1)! = 6$. Let $a := dim(J_4/J_4^2)$ and $b := dim(J_4^2/F_4^3)$. We know that $a + b + 6 = 55$ is valid. Thus, we want to determine b to calculate a. For calculating b we use part (i) of corollary 20 and theorem 23. The members of $\Pi_4^<$ are:
length 1: (1234);
length 2: $(1, 234), (12, 34), (13, 24), (14, 23), (123, 4), (124, 3), (134, 2)$;
length 3: $(1, 2, 34), (12, 3, 4), (13, 2, 4), (14, 2, 3), (1, 23, 4), (1, 24, 3)$;
length 4: $(1, 2, 3, 4)$.
The $c_{P,Q}$-values for $Q = (1, 2, 3, 4)$ are $4 \cdot 2 = 8$ and $3 \cdot 1 = 3$ where $4 - l(P) = 2$ and $(1, 2, 3, 4) < P$. The $c_{P,Q}$-values for $P = (1234)$ are $6x2 = 12$ where $l(Q) - 1 = 2$ and $Q < (1234)$. In addition, $b = 11 + 12 = 23$. Thus, $a = 55 - 6 - 23 = 26$.◇

Now we turn our focus to the ascending central chain of $1 + rad(K\Pi_n)$. Again, we want to determine the exponents of the factors alongside this chain. The next lemma describes some properties of the ascending central chain in a more general context.

Lemma 8 *Let K be a field and A a finite-dimensional associative unitary solvable K-algebra splitting over K such that a self-centralizing radical complement T exists which possesses a basis $\{e_1, ..., e_n\}$ consisting of pairwise orthogonal idempotents. The following statements are valid:*

(i) *Let $k \in \mathbb{N}$. The k-th member $Z_k(rad(A)^\circ)$ of the ascending central chain of $rad(A)^\circ$ is identical to $\sum\limits_{i \neq j=1, e_i Z_k(rad(A)^\circ)e_j \neq 0}^{n} e_i Z_k(rad(A)^\circ)e_j$.*

(ii) *Let $k \in \mathbb{N}$. Each member $Z_k(rad(A)^\circ)$ of the ascending central chain of $rad(A)^\circ$ is an ideal of the associative K-algebra A, $Z_{k+1}(rad(A)^\circ) \cdot rad(A), rad(A) \cdot Z_{k+1}(rad(A)^\circ) \subseteq Z_k(rad(A)^\circ)$ are valid and $Z_{k+1}(rad(A)^\circ)/Z_k(rad(A)^\circ)$ is an associative zero-algebra.*

(iii) *The center $Z(rad(A))$ of the nilradical is a zero-ideal of A which annihilates $rad(A)$.*

(iv) *If K is finite of characteristic p, then all factors alongside the ascending central chain of $1 + rad(A)$ are elementary-p-abelian.*

Proof: ad(i)-(iii): We start the proof with the case $k = 1$. Hence, $Z_k(rad(A)^\circ) = Z(rad(A)^\circ) = Z(rad(A))$ is the center of $rad(A)$. As done within a standard proof of the two-sided Pierce decomposition we use $1 = \sum\limits_{i=1}^{n} e_i$ and $Z(rad(A)) = 1 \cdot Z(rad(A)) \cdot 1$ to obtain that this center is contained in $\sum\limits_{i,j=1, e_i Z_k(rad(A)^\circ)e_j \neq 0}^{n} e_i Z_k(rad(A)^\circ)e_j$. For the set $e_i Z(rad(A))e_i$ $(1 \leq i \leq n)$ we use that T is self-centralizing: by lemma 5 we obtain that $e_i Z(rad(A))e_i$ is contained in $\langle e_i \rangle_K$. In addition, this space is contained in the ideal $rad(A)$. Thus, $e_i Z(rad(A))e_i$ is the zero-space. Let $i \neq j \in \underline{N}$, and $r \in rad(A)$. We calculate for an element $z \in Z(rad(A))$: $(e_i z e_j)r = e_i z(e_j r) = e_i(e_j r)z = 0$ and $r(e_i z e_j) = (r e_i)z e_j = z(r e_i)e_j) = 0$. Thus, we have proven all parts for the center of $rad(A)$.

Let us assume that for an element $k \in \mathbb{N}$ and for $Z_k(rad(A)^\circ)$ the parts (i) and (ii) are valid. We want to use an induction argument and the already proven statements for the center. In particular, $Z_k(rad(A)^\circ)$ is an ideal of A and we can define the associative algebra $B := A/Z_k(rad(A)^\circ)$. Hence, $rad(B) = rad(A)/Z_k(rad(A)^\circ)$ and $C := (T + Z_k(rad(A)^\circ))/Z_k(rad(A)^\circ)$ is a radical complement of $rad(B)$ in B which is isomorphic to T. We use lemma 5 and $(e_i + Z_k(rad(A)^\circ))B(e_i + Z_k(rad(A)^\circ)) = \langle e_i + Z_k(rad(A)^\circ \rangle_K$ $(1 \leq i \leq n)$ to obtain that C is self-centralizing. We use the proof for

128

the center to obtain both parts for the center of $rad(B)$. By definition, $Z(rad(B)) = Z_{k+1}(rad(A)^\circ)/Z_k(rad(A)^\circ)$ is valid. Now we use an induction argument and the theorem of homomorphism to derive that $Z_{k+1}(rad(A)^\circ)$ is an ideal of A and that $rad(A) \cdot Z_{k+1}(rad(A)^\circ)$ and $Z_{k+1}(rad(A)^\circ) \cdot rad(A)$ are contained in $Z_{k+1}(rad(A)^\circ)$. In particular, $Z_{k+1}(rad(A)^\circ)/Z_k(rad(A)^\circ)$ is a zero-algebra.

We have to prove part (i) for $Z_{k+1}(rad(A)^\circ)$. As done for the center at the beginning of this proof the only crucial part is the statement $e_i Z_{k+1}(rad(A)^\circ)e_j \subseteq Z_{k+1}(rad(A)^\circ)$ for all $i \neq j \in \underline{n}$. Let $i \neq j \in \underline{n}$, $z \in Z_{k+1}(rad(A)^\circ)$ and $r \in rad(A)$. We have to prove that $(e_i z e_j) \circ r$ is contained in $Z_k(rad(A)^\circ)$. But this is true because $Z_{k+1}(rad(A)^\circ)$ is and ideal of A which yields to $e_i z e_j \in Z_{k+1}(rad(A)^\circ)$. By definition, $(e_i z e_j) \circ r$ is contained in $Z_k(rad(A)^\circ)$.

ad(iv): For this part we present different proofs.

Variant 1: The center of $1 + rad(A)$ is exactly $1 + Z(rad(A))$. We use part (iii) to obtain that $Z(rad(A))$ is a zero-algebra. As proven within theorem 22 for each element $z \in Z(rad(A))$ we calculate $(1+z)^p = 1 + z^p = 1+0 = 1$. Thus, the center of $1 + rad(A)$ is elementary-p-abelian. We use a theorem of finite group that the exponent of the center is maximal within all factors alongside the ascending central chain (see [11], page 266). Hence, part (iv) is proven.

Variant 2: We use the theorem of Du (see [9]) to derive that the $(k+1)$-th center of $1 + rad(A)$ is exactly $1 + Z_{k+1}(rad(A)^\circ)$ for all $k \in \mathbb{N}$. Let $k \in \mathbb{N}$. In part (ii) we have proven that $Z_{k+1}(rad(A)^\circ)/Z_k(rad(A)^\circ)$ is a zero-algebra. Let $z \in Z_{k+1}(rad(A)^\circ)$. We obtain $(1+z)^p = 1 + z^p \in Z_k(rad(A)^\circ)$. Thus, all factors alongside the ascending central chain are elementary-p-abelian.

Variant 3: As proven within variant 1 the center of $1 + rad(A)$ is elementary-p-abelian. A theorem of radical algebras based on the theorem of Du (see theorem 5.2.5 in [30]) yields to the statement that all factors alongside the ascending central chain of $1 + rad(A)$ except the center are elementary-p-abelian.$\diamond$

As a direct consequence of the previous lemma 8, theorem 9 and corollary 1 we obtain:

Theorem 24 *Let K be a finite field of characteristic p and $n \in \mathbb{N}$. All factors alongside the ascending central chain of $1 + rad(K\Pi_n)$ are elementary-*

p-abelian.◇

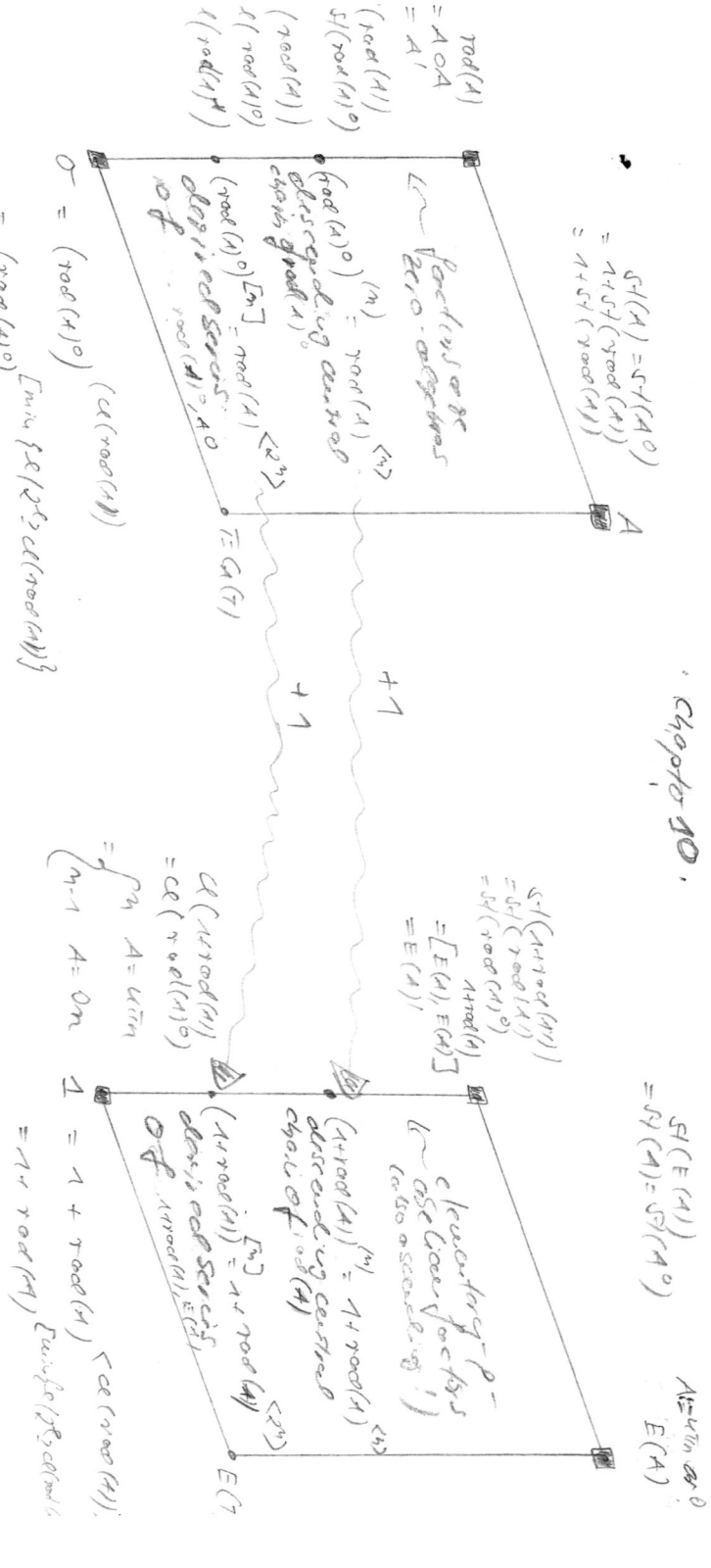

Algorithm 0:
E(A)

$S'(E(A)) = S'(A^0)$
$= S'(A) = S'(A^0)$

$S'(A) = S'(A^0)$
$= A^0 A$
$= A^i$

$rad(A)$
$= A^0 A$
$= A^i$

$(rad(A))$
$S'(rad(A)^0)$

$(rad(A))$
$S'(rad(A)^0)$

$S'(rad(A),A)$

$S'(A^{-1} + rad(A))$
$= S'(A + A^0)$
$= S'(rad(A))$
$A^{-1} + rad(A)$

Lu~ functions or the
zero categories

$(rad(A)^0)^{(m)}$: $rad(A)^{[m]}$
disregarding centres
chain grad A^0

$(rad(A)^0)[m] = rad(A)^{[2^2]}$
classi calSeries: $rad(A)^{[2^3]}$
of

$\sigma = (rad(A)^0)$ $(cl(rad(A)))$
classi calSeries: $rad(A)^0$
of $rad(A)^0$

$\sigma = (rad(A)^0)$ $[min \{ \varepsilon(2^3 cl(rad(A))\}]$

$= (rad(A)^0)$

$S'(A^{-1} + rad(A))$
$= S'(rad(A)^0)$
$= S'(rad(A))$
$= E(A)$

$(A + rad(A))^{(m)}$: $A + rad(A)^{[A]}$
descending centres
chain of (A)

$(A + rad(A))^{[m]} = A + rad(A)^{[2^3]}$
classi calSeries
of $A + rad(A), E(A)$

$cl(A + rad(A))$
$= cl(rad(A)^0$
of $A + rad(A)$

$cl(A + rad(A))$
$= 1 + rad(A)$ $\le cl(rad(A)),$
$\begin{cases} m & A = u^{-1}n \\ m-1 & A = 0n \end{cases}$

$1 = 1 + rad(A)$ $\le cl(rad(A)),$
$= 1 + rad(A)$ $[min \{ \varepsilon(2^3 cl rad... \}]$

— classes of n-factory of subfactors. L~g and exponents —

10.4 Open-ended questions

- Let K be a field and $n \in \mathbb{N}$. Determine the ascending central chain of $rad(K\Pi_n)^\circ$!

- Let K be a field and M a finite idempotent monoid. Is it possible to transfer the analysis and results of this chapter to KM?

- Does a general connection exist between the chain of derivations and the class of solvability for an associative unitary algebra, its associated Lie algebra and its group of units?

- Determine the ascending central chain of $1 + rad(K\Pi_n)$ and the size of the factors alongside the chain for a finite field.

- Let K be a field, $n \in \mathbb{N}$ and $k \in \underline{n-1}_{|}$. Are the dimensions of $rad(K\Pi_n)^{<k>}/rad(K\Pi_n)^{<k+1>}$ decreasing from top to bottom? What about the ascending central chain?

10.5 Exercises

Exercise 132 Let $n \in \mathbb{N}$. Prove that for each $s \in \underline{n+1}_{|}$ an element S in $\Pi_{n+1}^<$ of length s exists.

Exercise 133 Let K be a finite field of characteristic p. Investigate the structure of $1 + rad(K\Pi_n)$ for $n \le 5$ alongside the descending and ascending central chains and alongside the chain of derived subgroups! What is the influence of p and of the order of K?

Exercise 134 Let A be a nilpotent associative algebra of nilindex $n \in \mathbb{N}$. True or false: The chain of powers $A, A^2, ..., A^{n-1}, A^n = 0$ is the fastest chain $A = S_1, ..., 0 = S_r$ such that all factors alongside the chain are zero-algebras (with zero-multiplication).

Exercise 135 Let K be a field and $n \in \mathbb{N}$. For $n = 17, \cdots, 32$ add the classes of solvability for $1 + rad(K\Pi_n)$, $(K\Pi_n)^\circ$ and $E(K\Pi_n)$ to table 12.

Exercise 136 Let A be an associative unitary K-algebra and r, s nilpotent elements of class 2 of A such that $rsr - srs = 0$ is valid. Prove that $1+r, 1+s$ are units of A, sr is nilpotent of class 2 and $[1+s, 1+r] = 1 + s \circ r$ is valid. What is the inverse of $[1+s, 1+r]$?

Exercise 137 *Let K be a field of characteristic 0 and $n \in \mathbb{N}$. For $n = 17, \cdots, 32$ add the classes of solvability of $1 + rad(D_n)$, $(D_n)^\circ$ and $E(D_n)$ to table 13.*

Exercise 138 *For calculating the class of solvability and nilpotency a minimal $n \in \mathbb{N}$ must be determined for which 2^n is at least a constant value c. Express the statement by suing the logarithm-function and the ceiling-function.*

Exercise 139 *Within this chapter we calculate for an element $n \in \mathbb{N}$ and a field K for D_n and for $K\Pi_n$ classes of nilpotency and solvability. Do the difference $st(K\Pi_n) - st(D_n)$ and the fraction $\frac{st(K\Pi_n)}{st(D_n)}$ converge? What is the answer for the associated Lie algebra and the group of units of $K\Pi_n$ and D_n?*

Exercise 140 *Let K be a field of characteristic 0. Determine the ascending central chains of D_{35}, $(D_{35})^\circ$ and $E(D_{35})$!*

Exercise 141 *Let K be a field of characteristic 0. Determine the chain of derivations of $rad(D_{35})$, $rad(D_{35})^\circ$ and $1 + rad(D_{35})$!*

Exercise 142 *Let K be a field of characteristic 0. Determine the descending central chains of $1 + rad(D_{35})$ and $rad(D_{35})^\circ$!*

Exercise 143 *Let K be a field. Determine the chain of derivations for $K\Pi_{35}$, $K\Pi_{35}^\circ$ and $E(K\Pi_{35})$!*

Exercise 144 *Let K be a field. Determine the chain of derivations for $rad(K\Pi_{35})$, $rad(K\Pi_{35})^\circ$ and $1 + rad(K\Pi_{35})$!*

Exercise 145 *Let K be a field. Determine the descending central chains of $1 + rad(K\Pi_{35})$ and $rad(K\Pi_{35})^\circ$!*

Exercise 146 *Let K be a field and $n \in \mathbb{N}$. In what way is it possible do transfer the results and the analysis of this chapter to the algebra of lower triangular matrices of $K^{n \times n}$? (Tip: A sum-product-lemma is contained in [22].)*

Exercise 147 *Let K be a field and $n \in \underline{3}$. Determine the ascending central chains of $rad(K\Pi_n)^\circ$ and $1 + rad(K\Pi_n)$. (Tip: Theorem of Xiankun Du (see [9]))*

Exercise 148 *Let K be a field of characteristic zero and $n \in \underline{3}$. Determine the ascending central chains of $rad(D_n)^\circ$ and $1 + rad(D_n)$. (Tip: Theorem of Xiankun Du (see [9]))*

Chapter 11

The Lie nilradical and the Fitting subgroup

Within this chapter we focus on the connection of the Fitting subgroup and the nilradical motivated by a similar analysis done for Carter subgroups and Cartan subalgebras within the thesis of Thorsten Bauer in [3]:

- definition of a generalized Jordan[1] decomposition and application to the adjoint representation

- determining the nilradical of the Lie algebra associated to a finite-dimensional associative unitary solvable algebra possessing a separable radical factor algebra

- determining the nilradical of D_n° and $K\Pi_n^\circ$

- determining the Fitting[2] subgroup of the group of units of a finite-dimensional associative unitary solvable algebra possessing a separable radical factor algebra

- analyzing the connection of the Fitting subgroup and the Lie nilradical

- determining the Fitting subgroup of $E(D_n)$ and $E(K\Pi_n)$.

---

[1]Marie Ennemond Camille Jordan

[2]Hans Fitting

11.1 The nilradical of the Lie algebra associated to an associative algebra

Definitions 13 If L is a K-Lie algebra and $l \in L$, then we define the multiplication (also called adjoint representation) with l by $ad(l) : L \longrightarrow L, x \mapsto xl$. As done within definition 5.2.1 in [22] we call a polynomial fully-separable if it is semisimple and separable. If A is an associative unitary K-algebra, then an element $a \in A$ is called fully-separable if its minimal polynomial $min_{a,K}$ is fully-separable. As done within definition 5.1.4.1 in [22] a pair $(r; s) \in A \times A$ is called a (generalized) Jordan decomposition of $a \in A$, if $a = r + s$, r and s commute, r is nilpotent and s is fully-separable.$\diamond$

Lemma 9 *Let K be a field, A an associative unitary finite-dimensional K-algebra and $a, r, s \in A$. If $(r; s)$ is a (generalized) Jordan decomposition of a, then $(ad(r); ad(s))$ is one of $ad(a)$.*

Proof: *Step 1*: At first, we remark that a, $\lambda(a)$ and $\rho(a)$ possess the same minimal polynomial and $ad(a) = ad(r + s) = ad(r) + ad(s)$ is valid.

Step 2: We prove that $ad(r)$ and $ad(s)$ commute: A is associative. Thus, for all $x, y \in A$ the functions $\lambda(x)$ and $\rho(y)$ commute. r and s commute, and so do the functions $\lambda(r)$ and $\lambda(s)$ resp. $\rho(r)$ and $\rho(s)$. We calculate:

$$
\begin{aligned}
& ad(r)ad(s) \\
= \ & (\rho(r) - \lambda(r))(\rho(s) - \lambda(s)) \\
= \ & \rho(r)\rho(s) - \rho(r)\lambda(s) - \lambda(r)\rho(s) + \lambda(r)\lambda(s) \\
= \ & \rho(s)\rho(r) - \rho(s)\lambda(r) - \lambda(s)\rho(r) + \lambda(s)\lambda(r) \\
= \ & (\rho(s) - \lambda(s))(\rho(r) - \lambda(r)) \\
= \ & ad(s)ad(r).
\end{aligned}
$$

Step 3: Based on *Step 1* the functions $\rho(r)$ and $\lambda(r)$ are nilpotent, if r is nilpotent. A is associative, and thus the nilpotent endomorphism $\lambda(r)$ and $\rho(r)$ commute. Hence, their algebra-span is commutative and also nilpotent based on proposition 5 in [21]. We conclude that $ad(r)$ is nilpotent as difference of $\rho(r)$ and $\lambda(r)$.

Step 4: We use *Step 1*: the functions $\rho(s)$ and $\lambda(s)$ are fully-separable if s is fully-separable. A is associative, and thus the fully-separable endomorphism $\lambda(s)$ and $\rho(s)$ commute. Hence, their algebra-span is commutative and also fully-separable based on theorem 5.2.6 in [22]. We conclude that $ad(s)$ is fully-separable as difference of $\rho(s)$ and $\lambda(s)$.$\diamond$

Remark 20 The nilradical is the unique maximal nilpotent ideal of a Lie algebra (if it is existing). By a theorem of Fitting the sum of finite many nilpotent ideals is again nilpotent. If the Lie algebra is finite-dimensional, then the nilradical exists.

Let K be a field, A an associative unitary finite-dimensional solvable K-algebra possessing a separable radical factor algebra. We use [22] to derive that the center of A possesses a separable radical factor algebra, too, and for every radical complement T of A is $T \cap Z(A)$ the unique radical complement Z of $rad(Z(A))$ in $Z(A)$. It is identical to the intersection of all radical complements of $rad(A)$ in A.

Theorem 25 *Let K be a field, A an associative unitary finite-dimensional solvable K-algebra possessing a separable factor algebra by the nilradical and Z the unique radical complement of the center of A. The set $rad(A) \oplus Z$ is the nilradical and the unique maximal nilpotent Lie subalgebra of A° containing $rad(A)$.*

Proof: The Lie-ideals $rad(A)$ and Z of A° are nilpotent, and thus their sum is nilpotent, too. Hence, the Lie-nilpotent ideal $rad(A) \oplus Z$ is contained in the nilradical of A°. Let N be the nilradical of A° and $n \in N$. Based on the theorem of Wedderburn-Malcev a radical complement T exists. In addition, $r \in rad(A)$ and $t \in T$ exist such that $n = r + t$ is valid. $rad(A)$ is contained in N. Hence, $t \in N$ is true. N is a nilpotent Lie algebra, and thus $ad(t)$ is a nilpotent endomorphism of N. In particular, $ad(t)_{|rad(A)}$ is a nilpotent endomorphism of $rad(A)$. $A/rad(A)$ is separable and commutative, and we use theorem 5.3.1 in [22] to derive that every element of T is fully-separable. Based on lemma 9 we deduce that $ad(t)$ is fully-separable. In particular, $ad(t)_{|rad(A)}$ is fully-separable. Hence, $ad(t)_{|rad(A)}$ is nilpotent and fully-separable. We use lemma 9 and derive $ad(t)_{|rad(A)} = 0$. Thus, t centralizes $rad(A)$. T is commutative, and thus $t \in T \cap Z(A)$ is valid. This subalgebra is – based on theorem 5.1.4 in [22] – identical to Z.

The add-on is valid because every subalgebra containing $rad(A)$ is – because of $A \circ A \subseteq rad(A)$ – an ideal of A°.◇

Corollary 16 *Let K be a field and $n \in \mathbb{N}$. The following statements are valid:*

(i) The nilradical of $(K\Pi_n)^\circ$ is exactly $rad(K\Pi_n) \oplus K \cdot 1$.

(ii) For $char(K) = 0$ the nilradical of $(D_n)^\circ$ is exactly $rad(D_n) \oplus Z(D_n)$.

Proof: ad(i): see theorem 25 and theorem 13

ad(ii): see theorem 25 and theorem 3.6 in [3].◇

11.2 The Fitting subgroup of the group of units of an associative solvable algebra

The Fitting subgroup is the unique maximal nilpotent normal subgroup of a group (if it is existing). Based on a theorem of Fitting finite products of nilpotent normal subgroups are nilpotent. In particular, the Fitting subgroup exists for finite groups. Within the next theorem we use again remark 20.

Theorem 26 *Let K be a field, A an associative finite-dimensional unitary solvable K-algebra possessing a separable factor algebra by the nilradical and Z the radical complement of the center of A. The Fitting subgroup of $E(A)$ is the group of units of the nilradical of A°: $(1+rad(A)) \times E(Z)$. The second Fitting subgroup is the whole group of units.*

Proof: Based on theorem 25 is $rad(A) \oplus Z$ the nilradical of A°. $rad(A)$ is a nilpotent ideal and Z is central in A. Based on proposition 1.1.8 in [23] the group of units of the nilradical is the direct product of the nilpotent normal subgroup $1 + rad(A)$ and the normal subgroup $E(Z)$. Thus, the group of units of the nilradical of A° is a nilpotent normal subgroup of $E(A)$.

Let N be a nilpotent normal subgroup of $E(A)$. $1 + rad(A)$ is a nilpotent normal subgroup. We use a theorem of Hans Fitting to deduce the nilpotency of $N(1+rad(A))$. Based on lemma 2 in [4] the set $\langle (1+rad(A))N \rangle_K$ is Lie-nilpotent and contains $rad(A)$. Theorem 25 lets us deduce that $\langle N \rangle_K \subseteq \langle (1 + rad(A))N \rangle_K \subseteq rad(A) \oplus Z$ is valid. In particular, $N \leq E(\langle N \rangle_K) \leq E(\langle (1 + rad(A))N \rangle_K) \leq E(rad(A) \oplus Z)$ is true. The factor group by the Fitting subgroup is abelian, and thus the second Fitting subgroup is the whole group of units.⋄

Corollary 17 *Let K be a field and $n \in \mathbb{N}$. The following statements are valid:*

(i) The Fitting subgroup of $E(K\Pi_n)$ is the group of units of the nilradical of $(K\Pi_n)^\circ$.

(ii) If K possesses exactly 2 elements, then the Fitting subgroup of $E(K\Pi_n)$ is exactly $1 + rad(K\Pi_n)$.

(iii) If K possesses at least 3 elements, then the Fitting subgroup of $E(K\Pi_n)$ is exactly $(1 + rad(K\Pi_n)) \times ((K \setminus \{0\}) \cdot 1)$.

Proof: see theorem 26 and corollary 16.⋄

Corollary 18 *Let K be a field, $char(K) = 0$ and $n \in \mathbb{N}$. The following statements are valid:*

    (i) *The Fitting subgroup of $E(D_n)$ is the group of units of the nilradical of $(D_n)^\circ$.*

    (ii) *The Fitting subgroup of $E(D_n)$ ist exactly $(1 + rad(D_n)) \times E(Z(D_n))$.*

Proof: see theorem 26 and corollary 16.◇

T. Bauer has analyzed within his thesis [3] the dimension of the center of D_n. The value is depending purely on n, and the center is semi-simple. Within the exercises the reader may analyze this topic based on his thesis.

Chapter 11.

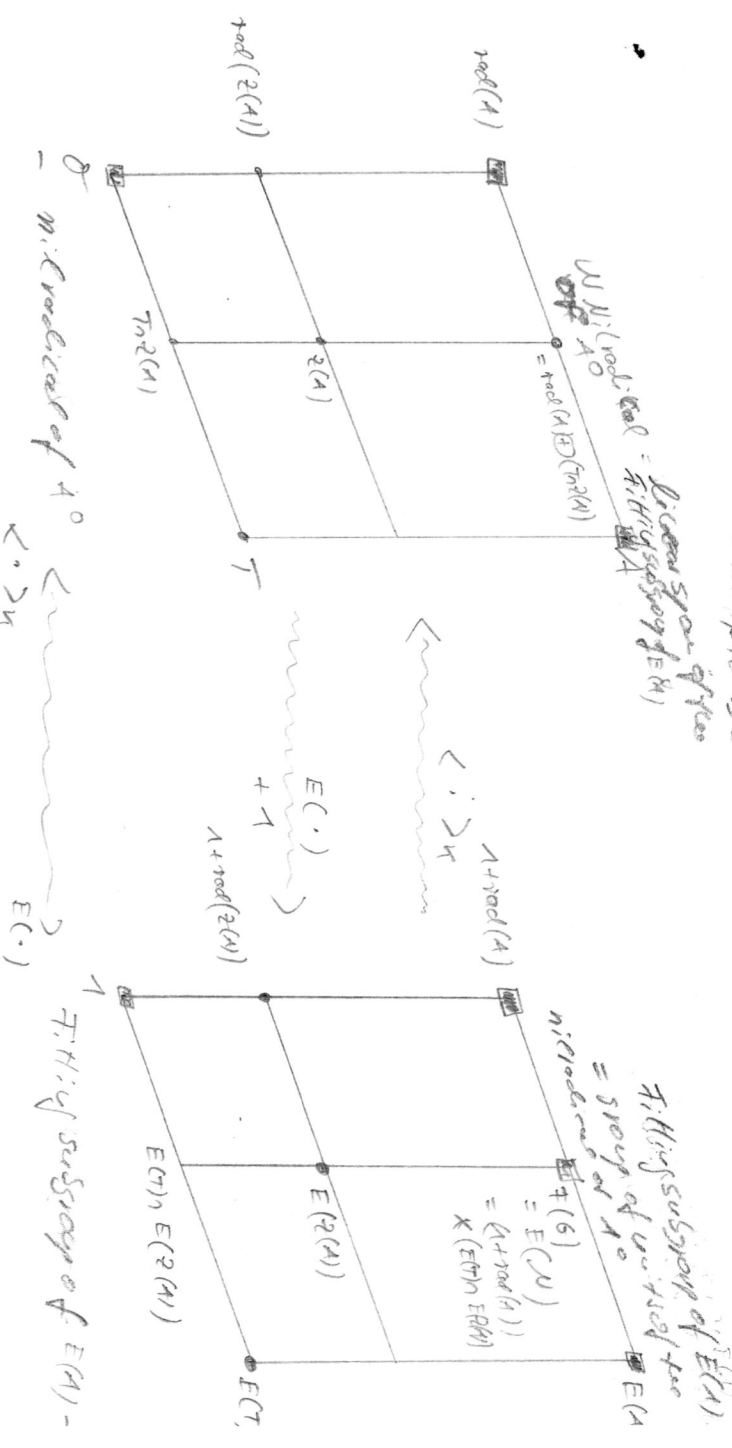

rad(A)

rad(Z(A))

W : Nilpoticeal = Lineare span of the
= rad(A) ⊕ rad(Z(A))
Fitting subgroup of E(A)

= rad(A) ⊕ (rad(A))

Z(A)

Tn Z(A)

T

Fitting subgroup of E(A).
= group of unities of E(A)
nilpotrad =

⟨ ⟩_T

E(·)
+ 1

1 + rad(Z(A))

1 + rad(A)

𝒯(6)
= E(U)
= (1 + rad(A))
× (E(T) ∩ E(A))

Fitting subgroup of E(A) =

E(T)∩E(Z(A))

E(Z(A))

E(T)

E(A

𝒪

M. Crocias of 1° ⟨⟶⟩_T   E(·)

... für Dn and UTn is the
zentra contained in T :

W(Dn°) = rad(Dn) ⊕ Z(Dn) ; 𝒯(E(Dn)) = (1+rad(Dn))×𝒯(⊆Z(Dn)),

W(UTn°) = rad(UTn)⊕Z(UTn) ; 𝒯(E(UTn)) = (1+rad(UTn))×E(Z(UTn))

11.3 Open-ended questions

- Let K be a field and M a finite idempotent monoid. What is the nilradical of KM°? What is the Fitting subgroup of $E(KM)$?

- Let K be a field and A a finite-dimensional associative unitary K-algebra. What is the nilradical of A°? What is the Fitting subgroup of $E(A)$? What is the connection between these two structures?

- Let K be a (finite) field and $n \in \mathbb{N}$. What is the[3] Frattini-subgroup of $E(K\Pi_n)$?

- What is the answer of the previous question in context of idempotent monoid algebra?

- What is the answer of the previous question in context of finite-dimensional associative (solvable) algebras over fields?

11.4 Exercises

Exercise 149 *Let K be a field. What is the nilradical of $(K\Pi_3)^\circ$? What is its connection to the Fitting subgroup of $E(K\Pi_3)$?*

Exercise 150 *Let K be a field. What is the Fitting subgroup of $E(K\Pi_3)$? What is its connection to the nilradical of $(K\Pi_3)^\circ$?*

Exercise 151 *Let K be a field and $a := 1 + (12, 3) - (13, 2) \in K\Pi_3$. Determine the Jordan decomposition of $ad(a)$.*

Exercise 152 *Let K be a field, A an associative unitary K-algebra and e an idempotent of A. Determine the Jordan decomposition $ad(e)$.*

Exercise 153 *Let K be a field. Does $e_{(12,3)} + e_{(3,12)}$ possess a Jordan decomposition in $K\Pi_3$?*

Exercise 154 *Prove lemma 9!*

Exercise 155 *Let A be an associative K-algebra and $(r; s)$ a Jordan decomposition of $r + s$. Prove that $(r\rho; s\rho)$ and $(r\lambda; s\lambda)$ are Jordan decompositions.*

---

[3]Giovanni Frattini

140

Exercise 156 *Let A be an associative K-algebra, $(r;s)$, (u,v) Jordan decompositions and $k \in K$. Are $(kr;ks)$, $(r+u;s+v)$ and $(ru;sv)$ Jordan decompositions? Analyze under which assumptions these pairs are Jordan decompositions!*

Exercise 157 *Let $n \in \mathbb{N}$ and K be a field. Determine the nilradical of the associated Lie algebra and the Fitting subgroup of the group of units of the algebra of lower triangular matrices of $K^{n \times n}$.*

Exercise 158 *Let $n \in \mathbb{N}$ and K be a field of characteristic zero. Study the thesis of Thorsten Bauer (see [3]) and determine the center of D_n, the nilradical of D_n° and the Fitting subgroup of $E(D_n)$.*

Exercise 159 *Let K be a field. Calculate the Jordan decomposition of the matrices $\begin{pmatrix} 1 & 2 & 3 \\ 0 & 1 & 2 \\ 0 & 0 & 1 \end{pmatrix}$ and $\begin{pmatrix} 1 & 2 \\ 1 & 1 \end{pmatrix}$. (Tip: see [22]!)*

Chapter 12

Semisimple left and right ideals

Within this chapter we analyze semisimple left and right ideals (also for the determination of the anti-automorphism of D_n and $K\Pi_n$ in the next chapter). The following topics are of special interest for finite-dimensional associative solvable K-algebras splitting over K and possessing self-centralizing radical complement:

- definition of left-sided, right-sided and two-sided Pierce-orthogonal elements

- description of maximal semisimple left and right ideals by using Pierce-orthogonal elements

- conjugation of theses maximal semisimple left and right ideals

- description of simple left and right ideals by using Pierce-orthogonal elements

- consequences for D_n and $K\Pi_n$.

12.1 Semisimple left and right ideals and Pierce-orthogonality in solvable associative algebras

Definitions 14 Let A be an associative K-algebra, $e_1, \cdots, e_n$ pairwise orthogonal idempotents of A such that their sum is 1_A and let $i \in \underline{n}$. We call e_i left- resp. right-sided Pierce-orthogonal if for all $j \in \underline{n} \setminus \{i\}$ the Pierce component $e_i A e_j$ resp. $e_j A e_i$ is the zero-space. e_i is called two-sided Pierce-orthogonal if e_i is left- and right-sided Pierce-orthogonal.$\diamond$

Proposition 17 *Let A be an associative unitary finite-dimensional solvable K-algebra with splitting field K, and let $e_1, \cdots, e_n$ pairwise orthogonal idempotents of A such that their K-span generates a self-centralizing radical complement and $i \in \underline{n}$. The following statements are valid:*

(i) *e_i is left-sided Pierce-orthogonal if and only if the right principal ideal $e_i A$ is one-dimensional (and thus spanned by e_i).*

(ii) *e_i is right-sided Pierce-orthogonal if and only if the left principal ideal $A e_i$ is one-dimensional (and thus spanned by e_i).*

(iii) *e_i is two-sided Pierce-orthogonal if and only if the principal ideal $A e_i A$ is one-dimensional (and thus spanned by e_i).*

Proof: ad(i): We use the right-sided Pierce-decomposition to derive $e_i A = \bigoplus_{j=1}^{n} e_i A e_j$. Hence, e_i is left-sided Pierce-orthogonal if and only if $e_i A = e_i A e_i$ is valid. Based on lemma 5 the Pierce-component $e_i A e_i$ is one-dimensional.

ad(ii): apply part (i) to the algebra A^-.

ad(iii): This part is – by using parts (i) and (ii) – equivalent to the identity $A e_i = \langle e_i \rangle_K = e_i A$. The principal ideal $A e_i A$ is one-dimensional. If it is one-dimensional, then $A e_i$ and $e_i A$ are one-dimensional because they are contained in $A e_i A$.◇

Proposition 18 *Let K be a field and A an associative finite-dimensional solvable K-algebra possessing a separable radical factor algebra. Every semisimple ideal is contained in the unique radical complement of the center of A.*
In particular, every two-sided Pierce-orthogonal element is central, if A splits over K and possesses a self-centralizing radical complement.

Proof: Let I be a semisimple ideal of A. Based on theorem 3.3.2 in [22] the ideal I is separable, and by using corollary 2.3.7 in [22] it is contained in a radical complement of A. We use corollary 2.3.7 in [22] to derive that all radical complements are conjugated under the group $(rad(A); \star)$. I is an ideal, and thus I is contained in the intersection of all radical complements. Corollary 5.1.5 in [22] is used to prove that this intersection is exactly the radical complement of the center of A. The add-on is a consequence of applying proposition 17, part (iii) on the ideal $A e_i A$.◇

Theorem 27 *Let A be an associative unitary finite-dimensional solvable K-algebra with splitting field K, and let $e_1, \cdots, e_n$ pairwise orthogonal idempotents of A such that their K-span generates a self-centralizing radical complement. The following statements are valid:*

(i) *$\langle e_i \mid i \in \underline{n}, e_i \; right-sided \, Pierce-orthogonal \rangle_K$ is a maximal semisimple left ideal of A.*

(ii) *All maximal semisimple left ideals of A are conjugated under $1 + rad(A)$.*

(iii) *Every semisimple left ideal of A is contained – modulo conjugation under $1 + rad(A)$ – in $\langle e_i \mid i \in \underline{n}, e_i \; right-sided Pierce-orthogonal \rangle_K$.*

(iv) *$\langle e_i \mid i \in \underline{n}, e_i \; left-sided \, Pierce-orthogonal \rangle_K$ is a maximal semisimple right ideal of A.*

(v) *All maximal semisimple right ideals of A are conjugated under $1 + rad(A)$.*

(vi) *Every semisimple right ideal of A is contained – modulo conjugation under $1 + rad(A)$ – in $\langle e_i \mid i \in \underline{n}, e_i \; left-sided Pierce-orthogonal \rangle_K$.*

(vii) *Every semisimple ideal of A is contained in the unique maximal semisimple ideal $\langle e_i \mid i \in \underline{n}, e_i \; two-sided Pierce-orthogonal \rangle_K$. This unique ideal is contained in the radical complement of the center of A and is the intersection of $\langle e_i \mid i \in \underline{n}, e_i \; left-sided Pierce-orthogonal \rangle_K$ and $\langle e_i \mid i \in \underline{n}, e_i \; right-sided Pierce-orthogonal \rangle_K$.*

Proof: ad(i)-(iii): At first we prove that the subalgebra $\hat{L} := \langle e_i \mid i \in \underline{n}, e_i \; right-sided Pierce-orthogonal \rangle_K$ is a semisimple left ideal. $\hat{L}$ is contained in the commutative radical complement $\langle e_1, \cdots, e_n \rangle_K$. Based on theorem 3.3.2 in [22] the set $\hat{L}$ is semisimple. Proposition 17 lets us derive that $A\hat{L} = \hat{L}$ is valid, and thus $\hat{L}$ is a semisimple left ideal of A.

Let L be a semisimple left ideal of A. Based on theorem 3.3.2 in [22] the subalgebra L is separable, and we can use corollary 2.3.7 to deduct that L is – modulo conjugation with $1 + rad(A)$ – contained in the radical complement $\langle e_1, \cdots e_n \rangle_K$. Again, by using theorem 3.3.2 in [22], a subset T of $\underline{n}$ exists such that $L = \langle e_t \mid t \in T \rangle_K$ is valid. Lemma 5 is used to derive $L = \bigoplus_{t \in T} e_t A e_t$. Let $t \in T$. We use the right-sided Pierce-decomposition to derive $A e_t = \bigoplus_{j=1}^{n} e_j A e_t$ and $A e_t \subseteq L$ because L is a left ideal. Thus, $e_j A e_t = 0$ is valid for all $j \neq t$.

144

ad(iv)-(vi): Apply parts (i)-(iii) to A^-.

ad(vii): Apply parts (i)-(vi) and proposition 18. Every ideal is fix under conjugation of units.◇

Theorem 28 *Let A be an associative unitary finite-dimensional solvable K-algebra with splitting field K, and let $e_1, \cdots, e_n$ pairwise orthogonal idempotents of A such that their K-span generates a self-centralizing radical complement. The following statements are valid:*

(i) *For every simple left ideal $L \neq 0$ of A is a right-sided Pierce-orthogonal e_i exists such that L – modulo conjugation under $1 + rad(A)$ – is identical to $\langle e_i \rangle_K$. These are all simple left ideals.*

(ii) *For every simple right ideal $R \neq 0$ of A is a left-sided Pierce-orthogonal e_i exists such that R – modulo conjugation under $1 + rad(A)$ – is identical to $\langle e_i \rangle_K$. These are all simple right ideals.*

(iii) *For every simple ideal $I \neq 0$ of A is a two-sided Pierce-orthogonal e_i exists such that I is identical to the central ideal $\langle e_i \rangle_K$. These are all simple ideals.*

Proof: use theorem 27 and 3.3.2 in [22] and proposition 17.◇

12.2 Consequences for $K\Pi_n$ and D_n

Proposition 19 *Let K be a field, $n \in \mathbb{N}$ and $P \in \Pi_n^<$. The following statements are valid:*

(i) *Only for $P = (\underline{n})$ is e_P right-sided Pierce-orthogonal.*

(ii) *Only for $P = (1, 2, \cdots, n)$ is e_P left-sided Pierce-orthogonal.*

(iii) *e_P is for $n \geq 2$ not two-sided Pierce-orthogonal.*

(iv) *e_P is for $n = 1$ and $P = (1)$ two-sided Pierce-orthogonal.*

Proof: ad(i): Based on page 22 stated before corollary 5.5 in [18] the identity $dim(K\Pi_n e_P) = l(P)!$ is valid. Thus, the left ideal $K\Pi_n e_P$ is of dimension one if P is of length 1. We conclude that $P = (\underline{n})$ is true.

ad(ii): Let $e_Q K\Pi_n e_P = 0$ be valid for all $Q \in \Pi_n^<$ such that $Q \neq P$ is true. The statements $(1, 2, \cdots, n) \in \Pi_n^<$ and $(1, 2, \cdots, n) \wedge Q = (1, 2, \cdots, n)$

are valid for all $Q \in \Pi_n$. Hence, $(1, 2, \cdots, n)$ is smaller than all elements of Π_n with respect to $<$. We use theorem 6.4 in [18] to deduce $P = (1, 2, \cdots, n)$. Let $Q \in \Pi_n^<$ and $e_Q K \Pi_n e_{(1,2,\cdots,n)} \neq 0$ be valid. Theorem 6.4 in [18] is used to derive $Q = Q \wedge (1, 2, \cdots, n)$. In particular, $l(Q) = n$ is true. Because of $Q \in \Pi_n^<$ we deduce $Q = (1, 2, \cdots, n)$. Thus, $e_Q K \Pi_n e_{(1,2,\cdots,n)} = 0$ is true for all $Q \in \Pi_n^<$ such that $Q \neq (1, 2, \cdots, n)$ is valid.

ad(iii)+(iv): This part is a direct consequence of parts (i) and (ii).$\diamond$

Theorem 29 *Let K be a field and $n \in \mathbb{N}$. The following statements are valid:*

(i) *The semisimple and simple left ideals $\neq 0$ of $K\Pi_n$ are the conjugates under $1 + rad(K\Pi_n)$ of $\langle e_{(n)} \rangle_K$.*

(ii) *The semisimple and simple right ideals $\neq 0$ of $K\Pi_n$ are the conjugates under $1 + rad(K\Pi_n)$ of $\langle e_{(1,2,\cdots,n)} \rangle_K$.*

(iii) *For $n \geq 2$ the algebra $K\Pi_n$ does not possess any semisimple ideals $\neq 0$.*

(iv) *For $n = 1$ the algebra $K\Pi_1$ is the only semisimple and simple ideal $\neq 0$ of $K\Pi_1$.*

Proof: use theorems 27, 28 and 9 and proposition 19 an.$\diamond$

Proposition 20 *Let K be a field of characteristic 0, $n \in \mathbb{N}$, D_n the Solomon algebra, $H_n := \langle \nu^p \mid p \vdash n \rangle_K$ spanned by pairwise orthogonal idempotents as constructed within lemma 3.4 in [3] and $p \vdash n$. The following statements are valid:*

(i) *ν^p is right-sided Pierce-orthogonal if and only if $k, d \in \underline{n}$ exist such that $p = d^k$ is valid.*

(ii) *ν^p is left-sided Pierce-orthogonal if and only if $d_1, d_2 \in \mathbb{N}_0$ exist such that $p = 2^{d_2} 1^{d_1}$ is valid.*

(iii) *ν^p is for uneven n two-sided Pierce-orthogonal if and only if $p = 1^n$ is valid.*

(iv) *ν^p is for even n two-sided Pierce-orthogonal if and only if $p = 1^n$ or $p = 2^{\frac{n}{2}}$ is valid.*

146

Proof: ad(i): Based on theorem 3.5 in [3] the dimension of the left ideal $D_n\nu^p$ is exactly the number of the associates of p. Lemma 9 and theorem 3.5 in [3] lets us deduce that H_n self-centralizing. We use proposition 17 to derive that ν^p is right-sided Pierce-orthogonal if and only if p possesses only one associate element. Thus, part (i) is proven.

ad(ii): Let $d_1, d_2 \in \mathbb{N}_0$ such that $p = 2^{d_2}1^{d_1}$ is valid. We assume that $\nu^p D_n \nu^r \neq 0$ is valid for a partition r of n. Based on theorem 3.5 in [3] a power-free decomposition q of p exists which is associated to r. The only power-free decomposition of 2 is 2 and of 1 is 1. Thus, q and p are identical. In every association-class only one partition exists, and hence $p = q = r$ is valid. We only have to prove that these are the only possible values for p. Let $p = p_1 \cdots p_k$ and $p_1 \geq 3$. Let r be a partition of n which is associated to $p_2 \cdots p_k(p_1 - 1)1$. We define $a := p_1 - 1$ and $b := 1$, and we derive $a \neq b$ (because of $p_1 \geq 3$) and $pab \cong r(a+b)$. Now we use theorem 3.5 in [3] to derive $\nu^p D_n \nu^r \neq 0$.

ad(iii)+(iv): use parts (i) and (ii).$\diamond$

Theorem 30 *Let K be a field of characteristic 0, $n \in \mathbb{N}$, D_n the Solomon algebra, $H_n := \langle \nu^p \mid p \vdash n \rangle_K$ spanned by pairwise orthogonal idempotents as constructed within lemma 3.4 in [3] and $p \vdash n$. The following statements are valid:*

(i) *$\langle \nu^p \mid \exists k, d \in \mathbb{N}_0 : p = d^k \rangle_K$ is a maximal semisimple left ideal of D_n.*

(ii) *All maximal semisimple left ideals of D_n are conjugated under $1 + rad(D_n)$.*

(iii) *Every semisimple left ideal of D_n is contained – modulo conjugation under $1 + rad(D_n)$ – in $\langle \nu^p \mid \exists k, d \in \mathbb{N}_0 : p = d^k \rangle_K$.*

(iv) *$\langle \nu^p \mid \exists d_1, d_2 \in \mathbb{N}_0 : p = 2^{d_2}1^{d_1} \rangle_K$ is a maximal semisimple right ideal of D_n.*

(v) *All maximal semisimple right ideals of D_n are conjugated under $1 + rad(D_n)$.*

(vi) *Every semisimple right ideal of D_n is contained – modulo conjugation under $1 + rad(D_n)$ – in $\langle \nu^p \mid \exists d_1, d_2 \in \mathbb{N}_0 : p = 2^{d_2}1^{d_1} \rangle_K$.*

(vii) *For uneven n the set $\langle \nu^{1^n} \rangle_K$ is the only semisimple ideal $\neq 0$ of D_n.*

(viii) *For even n the sets $\langle \nu^{1^n} \rangle_K$, $\langle \nu^{2^{\frac{n}{2}}} \rangle_K$ and $\langle \nu^{1^n}, \nu^{2^{\frac{n}{2}}} \rangle_K$ are the only semisimple ideals $\neq 0$ of D_n.*

Proof: Based on lemma 9 and theorem 3.5 in [3] the set H_n is self-centralizing. The proof is now a consequence of theorem 27 and proposition 20.⋄

Theorem 31 *Let K be a field of characteristic 0, $n \in \mathbb{N}$, D_n the Solomon algebra, $H_n := \langle \nu^p \mid p \vdash n \rangle_K$ spanned by pairwise orthogonal idempotents as constructed within lemma 3.4 in [3] and $p \vdash n$. The following statements are valid:*

(i) *For every simple left ideal $L \neq 0$ elements $k, d \in \mathbb{N}_0$ exists such that L – modulo conjugation with $1 + rad(D_n)$ – is exactly $\langle \nu^{d^k} \rangle_K$. These are all simple left ideals.*

(ii) *For every simple right ideal $R \neq 0$ elements $d_2, d_1 \in \mathbb{N}_0$ exists such that R – modulo conjugation with $1 + rad(D_n)$ – is exactly $\langle \nu^{2^{d_2} 1^{d_1}} \rangle_K$. These are all simple right ideals.*

(iii) *The simple ideals $\neq 0$ of D_n are – for even n – exactly $\langle \nu^{1^n} \rangle_K$ and $\langle \nu^{2^{\frac{n}{2}}} \rangle_K$.*

(iv) *For uneven n the set $\langle \nu^{1^n} \rangle_K$ is the only simple ideal $\neq 0$ of D_n.*

Proof: Based on lemma 9 and theorem 3.5 in [3] the set H_n is self-centralizing. The proof is now a consequence of the theorems 28 and 3.3.2 in [22] and of proposition 20.⋄

$K^{n \times n}$

e_{ij}, $e \cdot TI_n \longrightarrow$ Basis $\longrightarrow r^n$, $p + m$

$e_{ij} -$ left-sided $\longrightarrow$ Ruíce ortogonal $\longrightarrow r^2 \, d_1 \wedge d_1$

$e(t_1, t_2) -$ right-sided $\longrightarrow$ Ruíce ortogonal $\longrightarrow r \, d \, k$

$t -$ two-sided $\longrightarrow$ Ruíce ortogonal $\longrightarrow r \, d_1 \, a_2$

- Ruíce orthogonale $\in K^{n \times n}$
  in $K^{n \times n}$ $a_1 a_2$ O_n

ad(A) $\longrightarrow$ $A_{ee} = \langle e \cdot a \rangle_u$

single
left-ideal
$A_{ee} = \langle e a \rangle_u$

$\left(\begin{array}{c} \text{an equivalent way of semi-single} \\ \text{left-ideal} \end{array} \right)$

$= \langle e \, e \cdot \text{rect} \rangle_u$
  Ruíce ortogonal

an equivalent way of semi-single
weight left-ideal
$1 + \text{rod}(A)$

0 \quad an inner
  T \quad left-ideal
  single-ideal

an inner
  R two-sided-ideal
  single-right-ideal
  all categorical
  inner-ideal

$T = C_A(T)$
  $= \langle e_A \rangle \cdots e_n \rangle_u$
  $= k^m$

$\langle e_A \mid e_A \text{ two-sided} \rangle_u$
  Ruíce ortogonal

single
  left-ideal
  $e + A = \langle e_A \rangle_u$
  Ruíce ortogonal

$\langle e_A \mid e_A \text{ left-sided} \rangle_u$
  Ruíce ortogonal

$1 + \text{rod}(A)$
  $= \langle e_A \mid e_A \text{ left-sided} \rangle_u$
  Ruíce ortogonal

single
ideals

$\langle e_A \mid e_A \text{ two-sided} \rangle_u$
  Ruíce ortogonal

$A_{ee} + A = \langle e_A \rangle_u$

$I = L \cap R$

- covariance
- Socrate maximal single substructures in
  Socrate $0 C_i$ this associate optical radius
  for centralizing radical complexes
  for centralizing radical complexes

12.3 Open-ended questions

- Is it possible to transfer and extend the results of this chapter to Solomons algebra in positive characteristics?

- Is it possible to transfer and extend the results of this chapter to idempotent monoid algebras?

- Is it possible to transfer and extend the results of this chapter to in a more general context of associative algebras?

12.4 Exercises

Exercise 160 *Let A be an associative unitary K-algebra, R a right and L a left ideal of A and $r, s \in rad(A)$. The element $1 + r$ is an unit of A and the identities $L^{1+r} = L(1 + r)$ and $R^{1+r} = (1 + r')R$ are valid. On what terms is $L(1 + r) = (1 + s')R$ true? What is the answer in the case $r = s$?*

Exercise 161 *Determine the unordered set partitions of $\underline{6}$ and their number!*

Exercise 162 *Let K be a field. What is the dimension of $e_{(1234)}K\Pi_4$? (Tip: Study the proofs of this chapter!)*

Exercise 163 *True or false: Maximal semi-simple left and right ideals and semi-simple ideals are existing within associative algebras.*

Exercise 164 *Let K be a field. What is the dimension of $K\Pi_4 e_{(1,2,3,4)}$? (Tip: Study the proofs of this chapter!)*

Exercise 165 *Let K be a field. Determine the simple, semi-simple and separable left ideals of $K\Pi_3$ and $K\Pi_4$! What is the dimension of the maximal ones?*

Exercise 166 *Let K be a field. Determine the simple, semi-simple and separable right ideals of $K\Pi_3$ and $K\Pi_4$! What is the dimension of the maximal ones?*

Exercise 167 *Let K be a field. Determine the simple, semi-simple and separable ideals of $K\Pi_3$ and $K\Pi_4$! What is the dimension of the maximal ones?*

Exercise 168 *Let K be a field. Determine the left- and right-sided and two-sided Pierce-orthogonal idempotents with respect to the orthogonal idempotents $e_P, P \in \Pi_5^{\leq}$!*

Exercise 169 *Let K be a field of characteristic zero. Determine the left- and right-sided and two-sided Pierce-orthogonal idempotents with respect to the orthogonal idempotents $\nu^p, p \vdash 6$!*

Exercise 170 *Let K be a field of characteristic zero. Determine the simple, semi-simple and separable left ideals of D_3 and D_4! What is the dimension of the maximal ones?*

Exercise 171 *Let K be a field of characteristic zero. Determine the simple, semi-simple and separable right ideals of D_3 and D_4! What is the dimension of the maximal ones?*

Exercise 172 *Let K be a field of characteristic zero. Determine the simple, semi-simple and separable ideals of D_3 and D_4! What is the dimension of the maximal ones?*

Exercise 173 *Let K be a field and $n \in \mathbb{N}$. Transfer the results of this chapter to the algebra of lower triangular matrices of $K^{n \times n}$!*

Exercise 174 *Transfer the results of this chapter for semi-simple (one-sided) ideals to separable ones! (Tip: semi-simple = separable within solvable algebras!)*

Exercise 175 *Let K be a field, $n \in \mathbb{N}$ and $P \in \Pi_n$. Is e_p left-sided and/or right-sided Pierce-orthogonal?*

Exercise 176 *Let A be an associative K-algebra, $r \in rad(A)$, $e_1, \cdots, e_n$ pairwise orthogonal idempotents of A decomposing 1_A as sum and $i \in \underline{n}$. If e_i is left-sided/right-sided/two-sided Pierce-orthogonal, then $(e_i)^{1+r}$ has the same property with respect to the pairwise orthogonal idempotents $(e_1)^{1+r}, \cdots, (e_n)^{1+r}$, and their sum is again 1_A.*

Exercise 177 *Let A be an associative K-algebra, $r \in rad(A)$, $e_1, \cdots, e_n$ and $f_1, \cdots, f_n$ pairwise orthogonal idempotents of A such that both sums are 1_A and both K-spans for a (self-centralizing) radical complement. What is the connection between the radical complements and the idempotents? (Tip: theorem, Wedderburn-Malcev, uniqueness of decomposition, exercise 176)*

Exercise 178 *Let A be an associative right-artian K-algebra, I an ideal, L a left and R a right ideal of A. The following statements are valid:*

(i) $L + R$ is a subalgebra of A.

(ii) $L \cap R$ *is an ideal of* A.

(iii) $L/(L \cap R)$ *and* $L/(L \cap R)$ *are ideals of* $(L + R)/(L \cap R)$.

(iv) $L/(L \cap R)$ *and* $L/(L \cap R)$ *are direct within* $(L + R)/(L \cap R)$.

(v) *If* I *and* A/I *are semi-simple, then* A *is semi-simple, too, and vice-versa. (Tip: What is the nilradical of* A/I *resp. of* I*?)*

(vi) *Let* A *be finite-dimensional and solvable and* L, R *are semi-simple. Prove that* $L + R$ *is semi-simple.*

Remark: The last part can be applied to the maximal right and left ideals within theorem 27. The structural importance of these semi-simple substructures is not known to the author. Their intersection is the unique maximal semi-simple ideal of A. *The reader may analyze the importance of these statements within* $K\Pi_n$ *and* D_n!

Chapter 13

Anti-automorphism

Within this chapter we determine the anti-automorphism of D_n and $K\Pi_n$ and focus on the following topics:

- determination the dimension of the maximal semisimple left and right ideals of D_n

- proof that almost no anti-automorphism of D_n exist

- determination of the anti-automorphism of D_n in the remaining cases of n

- determination of the maximal dimension of projective indecomposable left and right ideals of $K\Pi_n$

- Proof that almost no anti-automorphism of $K\Pi_n$ exist

- determination of the anti-automorphism of $K\Pi_n$ in the remaining cases of n.

13.1 Dimensions of maximal semisimple left and right ideals of D_n

Definitions 15 Let A, B be K-algebras. By $Iso(A, B)$ (more exact $Iso_K(A, B)$) we denote the set of all K-algebra isomorphism between A and B. In particular, we use $Aut(A) := Iso(A, A)$ and $Ant(A) := Iso(A, A^-)$. The elements of $Aut(A)$ resp. of $Ant(A)$ are called automorphism resp. anti-automorphism of A. The inner automorphism of A are symbolized by $Inn(A)$.

For every $n \in \mathbb{N}$ let $\tau(n)$ be the number of divisors of n and C_n be the cyclic group of order n. $\diamond$

Corollary 19 *Let K be a field, $char(K) = 0$ and $n \in \mathbb{N}$.*

(i) All maximal semisimple left ideals of D_n are of dimension $\tau(n)$.

(ii) For even n all maximal semisimple right ideals of D_n are of dimension $\frac{n}{2} + 1$.

(iii) For uneven n all maximal semisimple right ideals of D_n are of dimension $\frac{n+1}{2}$.

(iv) All maximal semisimple ideals of D_n are of dimension 1.

(v) D_n is commutative if and only if $n = 1, 2$ is valid.

Proof: ad(i): Apply the parts (i) and (ii) of theorem 30.

ad(ii): For even n the partitions of n of shape $2^{d_2} 1^{d_1}$ with $d_2, d_1 \in \mathbb{N}_0$ are exactly

$$
\begin{array}{c}
1 \cdots 1 \\
21 \cdots 1 \\
221 \cdots 1 \\
\cdots \\
2 \cdots 211 \\
2 \cdots 2.
\end{array}
$$

Their number is exactly $\frac{n}{2} + 1$. Parts (iv) and (v) of theorem 30 are used to prove (ii).

ad(iii): For uneven n the partitions of n of shape $2^{d_2} 1^{d_1}$ with $d_2, d_1 \in \mathbb{N}_0$ are exactly

$$
\begin{array}{c}
1 \cdots 1 \\
21 \cdots 1 \\
221 \cdots 1 \\
\cdots \\
2 \cdots 2111 \\
2 \cdots 21.
\end{array}
$$

Their number is exactly $\frac{n+1}{2}$. Parts (iv) and (v) of theorem 30 are used to prove (iii).

ad(iv): Apply the parts (vii) and (viii) of theorem 30.

ad(v): We use lemma 5 and part (f) of theorem 3.5 in [3] to prove the

n	$\tau(n)$	$\frac{n}{2}+1$	$\frac{n+1}{2}$
1	1		1
2	2	2	
3	2		2
4	3	3	
5	2		3
6	4	4	
7	2		4
8	4	5	
9	3		5
10	4	6	.

Table 13.1: dimensions of maximal semisimple left and right ideals of D_n

existence of a self-centralizing radical complement. Thus, D_n is commutative if and only if D_n is semisimple. Based on theorem 3.3 in [3] the nilradical of D_n is zero if and only if every decompositions of n possesses exactly one associated element. This is possible only for $n = 1, 2.\diamond$

Example 9 Let $n \in \mathbb{N}$. We list – using parts (i), (ii) and (iii) of corollary 19 – the calculated dimensions for $n \leq 10$ within 13.1. The value $\frac{n}{2}+1$ resp. $\frac{n+1}{2}$ is filled only for even resp. uneven n. If $n = \prod_{i=1}^{s} p_i^{r_i}$ is the decomposition of n into primes, then the existence and uniqueness of such a decomposition lets us derive the formula $\tau(n) = \prod_{i=1}^{r_i}(r_i + 1).\diamond$

13.2 Antiautomorphism of D_n

Proposition 21 *Let K be a field, $char(K) = 0$ and $n \in \mathbb{N}$. For $n \leq 2$ resp. $n \in \mathbb{N}_{\geq 7} \cup \{5\}$ Solomons algebra D_n possesses one resp. no antiautomorphism.*

Proof: Every anti-automorphism sends a semisimple left resp- right ideal to a semisimple right resp. left ideal. Thus, the existence of an anti-automorphism yields to the statement that the maximal dimension of semisimple left and right ideals are identical (existence see 30). Because of corollary 19 we only need to prove that $\tau(n) < \frac{n+1}{2} \leq \frac{n}{2} + 1$ is valid. We use [15] to deduce $\tau(n) \leq \frac{7}{4}\sqrt[2]{n}$. On addition, for $n \geq 11$ the inequality $\frac{7}{4}\sqrt[2]{n} < \frac{n+1}{2}$ is true. For $n = 5, 7, 8, 9, 10$ the inequality $\tau(n) < \frac{n+1}{2}$ is valid based on

	3	21	111
3	1	1	0
21	0	1	0
111	0	0	1.

Table 13.2: Cartan matrix of D_3

example 9.

Let $n \leq 2$. The dimension of D_n is at most 2. D_n is unitary, and thus D_n is commutative. In this case an anti-automorphism is the identity $id_{D_n}.\diamond$

The parts (i) and (ii) of the following proposition are based on a correspondence with Thorsten Bauer:

Proposition 22 *Let K be a field, $char(K) = 0$, $n \in \mathbb{N}$, and D_n Solomons algebra with radical complement $H_n := \langle \nu^p \mid p \vdash n \rangle_K$ spanned of pairwise orthogonal idempotents as presented within lemma 3.4 in [3]. The following statements are valid:*

  (i) *D_3 possesses an anti-automorphism, namely the linearization α_3 of the function $\nu^3 \mapsto \nu^{21}, \nu^{21} \mapsto \nu^3, \nu^{111} \mapsto \nu^{111}, \nu^3 \nu_{21} \mapsto \nu^3 \nu^{21}$.*

  (ii) *D_4 possesses an anti-automorphism, namely the linearization α_4 of the function $\nu^4 \mapsto \nu^{211}, \nu^{31} \mapsto \nu^{31}, \nu^{211} \mapsto \nu^4, \nu^{22} \mapsto nu^{22}, \nu^{1111} \mapsto \nu^{1111}, \nu^4 \nu_{31} \mapsto \nu^{31} \nu_{211}, \nu^{31} \nu_{211} \mapsto \nu^4 \nu_{31}, \nu^4 \nu_{211} \mapsto \nu^4 \nu_{211}$.*

  (iii) *D_6 possesses no anti-automorphism.*

  (For the definition of the elements $\nu_p, p \vdash n$ and $\omega_q, q \models n$ see chapter 3 in [3].)

Proof: ad(i): We focus on the two-sided Pierce-decomposition related to the pairwise orthogonal idempotents $\nu^p, p \vdash n$. Based on theorem 3.5 in [3] only the following Pierce components are different from zero: $\nu^3 D_3 \nu^3 = \langle \nu^3 \rangle_K$, $\nu^{21} D_3 \nu^{21} = \langle \nu^{21} \rangle_K$, $\nu^{111} D_3 \nu^{111} = \langle \nu^{111} \rangle_K$ and $\nu^3 D_3 \nu^{21} = \langle \nu^3 \nu_{21} \rangle_K$. Thus, $B_3 := \{\nu^3, \nu^{21}, \nu^{111}, \nu^3 \nu_{21}\}$ is a K-basis of D_3. The Cartan matrix is presented within table 13.2. We only need to prove that for all $a, b \in B_3$ the identity $(ab)\alpha_3 = (b\alpha_3)(a\alpha_3)$ is valid. The corresponding calculations are placed within tables 13.3 and 13.4. The 'transposed matrix' of the second table is exactly the first table. The calculations use the orthogonality of the idempotents $\nu^p, p \vdash n$ and lemma 3.4 in [3]: For all $q \models n$ and $r \vdash n$ the statement $\omega_q \nu^r = \omega_q$ is valid, if q and r are associated. In the other case

$(ab)\alpha_3$	v^3	v^{21}	v^{111}	v^3v_{21}
v^3	$v^3\alpha_3 = v^{21}$	$0\alpha_3 = 0$	$0\alpha_3 = 0$	$v^3v_{21}\alpha_3 = v^3v_{21}$
v^{21}	$0\alpha_3 = 0$	$v^{21}\alpha_3 = v^3$	$0\alpha_3 = 0$	$0\alpha_3 = 0$
v^{111}	$0\alpha_3 = 0$	$0\alpha_3 = 0$	$v^{111}\alpha_3 = v^{111}$	$0\alpha_3 = 0$
v^3v_{21}	$0\alpha_3 = 0$	$v^3v_{21}\alpha_3 = v^3v_{21}$	$0\alpha_3 = 0$	$0\alpha_3 = 0$

Table 13.3: composition of the anti-automorphism α_3 of D_3, part 1

$(a\alpha_3)(b\alpha_3)$	v^{21}	v^3	v^{111}	v^3v_{21}
v^{21}	v^{21}	0	0	0
v^3	0	v^3	0	v^3v_{21}
v^{111}	0	0	v^{111}	0
v^3v_{21}	v^3v_{21}	0	0	0

Table 13.4: composition of the anti-automorphism α_3 von D_3, part 2

$\omega_q v^r = 0$. In addition, for all $p \models n$ the element ν_p is – modulo a factor of K – identical to ω_p. The corresponding calculations are summarized within tables 13.3 and 13.4.

ad(ii): As presented within part (i) we use theorem 3.5 in [3] to derive that the set $B_4 := \{\nu^4, \nu^{31}, \nu^{211}, \nu22, \nu^{1111}, \nu^4\nu_{31}, \nu^{31}\nu_{211}, \nu^4\nu_{211}\}$ is a K-basis of D_4 because the Cartan matrix is of the content presented within table 13.5. We only need to prove that for all $a, b \in B_4$ the equation $(ab)\alpha_4 = (b\alpha_4)(a\alpha_4)$ is valid. The corresponding calculations are presented within the following tables. The 'transpose' of the second table is exactly the first table. The calculations are based on the orthogonality of the idempotents $\nu^p, p \vdash n$ and lemma 3.4 in [3]: For all $q \models n$ and $r \vdash n$ the identity $\omega_q v^r = \omega_q$ is valid, if q and r are associated. In the other case the statement $\omega_q v^r$ is zero. In addition, for all $p \models n$ the element ν_p – modulo a factor of K – is identical to ω_p. We use the description of the Pierce components within theorem 3.5 in

	4	31	22	211	1111
4	1	1	0	1	0
31	0	1	0	1	0
22	0	0	1	0	0
211	0	0	0	1	0
1111	0	0	0	0	1.

Table 13.5: Cartan matrix of D_4

[3] to deduce that $\gamma := v^4 v_{31} v^{31} v_{211}$ is a fixpoint under α_4. The calculations are placed within tables 13.7 and 13.8.

ad(iii): If an anti-automorphism exists, then we can use the theorem of[123] Krull-Remak-Schmidt to deduce that the maximum of the set $\{dim_K(D_n \nu^p) \mid p \vdash n\}$ and the maximum of the set $\{dim_K(\nu^p D_n) \mid p \vdash n\}$ are identical: The sets are containing the dimensions of the projective indecomposable left resp. right ideals. We use theorem 3.5 in [3] to derive that for $p \vdash n$ the dimension of the left ideal $D_n \nu^p$ is exactly the number of associate elements of p. Let $p \vdash n$. The right ideal $\nu^p D_n$ is exactly $\bigoplus_{r \vdash n} \nu^p D_n \nu^r$ (see theorem 3.5 in [3]). The Pierce component $\nu^p D_n \nu^r$ is – based on theorem 3.5 in [3] – not zero if and only if a power-free decomposition q of p exists which is associated to r. Within table 13.6 the dimensions of the left ideal $D_6 \nu^p$ for $p \vdash 6$ are presented in the second column. In addition, we use X in the third column to mark if the Pierce component $\nu^6 D_6 \nu^p$ is for $p \vdash 6$ not equal to zero. The first column lists all partitions of 6. The content of this table lets us deduce that the maximal dimension of the left ideals $D_6 \nu^p$ is for $p \vdash 6$ exactly 6, but the maximal dimension of the right ideals $\nu^p D_6$ is at least 8. Thus, part (iii)

[1] Wolfgang Krull

[2] Robert Erich Remak

[3] Otto Juljewitsch Schmidt

$p \vdash 6$	6	51	42	33	321	411	222	3111	2211	21111	111111
$dim_K(D_6 \nu^p)$	1	2	2	1	6	3	1	4	6	5	1
$\nu^6 D_6$	X	X	X		X	X		X	X	X	

Table 13.6: D_6 contains no anti-automorphism

is proven.$\diamond$

The parts (ix) and (x) of the following proposition are based on another correspondence with Thorsten Bauer:

Proposition 23 *Let A be an algebra and $\beta \in Ant(A)$. The following statements are valid:*

(i) $Iso(A, A^-) = Iso(A^-, A)$

(ii) $Iso(A, A) = Iso(A^-, A^-)$

(iii) $Ant(A) = Ant(A^-)$

(iv) $Aut(A) = Aut(A^-)$

(v) $Ant(A)^{-1} = Ant(A)(= Ant(A^-))$

(vi) $Ant(A)Aut(A) \subseteq Ant(A)$

(vii) $Aut(A)Ant(A) \subseteq Ant(A)$

(viii) $Ant(A)Ant(A) \subseteq Aut(A)$

(ix) $Ant(A) = \beta Aut(A) = Aut(A)\beta$
If an anti-automorphism of A exists, then the number of anti-automorphism and of the automorphism of A are identical.

(x) $Aut(A) = \beta Ant(A) = Ant(A)\beta$

(xi) *A commutative $\iff id_A \in Ant(A) \iff Ant(A) = Aut(A)$*

Proof: ad(i): An easy calculation yields to the fact that $Iso(A, A^-)$ is contained in $Iso(A^-, A)$. By applying this inclusion with A^- instead of A, then part (i) is proven.

$(a\alpha_4)(b\alpha_4)$	v^{211}	v^{31}	v^4	v^{22}	v^{1111}	$v^{31}v_{211}$	v^4v_{231}	v^4v_{211}
v^{211}	v^{211}	0	0	0	0	0	0	0
v^{31}	0	v^{31}	0	0	0	0	0	0
v^4	0	0	v^4	0	0	$v^{31}v_{211}$	v^4v_{31}	v^4v_{211}
v^{22}	0	0	0	v^{22}	0	0	0	0
v^{1111}	0	0	0	0	v^{1111}	0	0	0
$v^{31}v_{211}$	$v^{31}v_{211}$	0	0	0	0	0	0	0
v^4v_{31}	0	v^4v_{31}	0	0	0	γ	0	0
v^4v_{211}	v^4v_{211}	0	0	0	0	0	0	0

Table 13.7: composition of the anti-automorphism α_4 of D_4, part 1

$(ab)\alpha_4$	v^4	v^{31}	v^{211}	v^{22}	v^{1111}	$v^4 v_{31}$	$v^{31}v_{211}$	$v^4 v_{211}$
v^4	v^{211}	$0\alpha_4 = 0$	$0\alpha_4 = 0$	$0\alpha_4 = 0$	$0\alpha_4 = 0$	$v^4 v_{31}\alpha_4 = v^{31}v_{211}$	$0\alpha_4 = 0$	$v^4 v_{211}\alpha_4 = v^4 v_2$
v^{31}	0	$v^{31}\alpha_4 = v^{31}$	$0\alpha_4 = 0$	$0\alpha_4 = 0$	$0\alpha_4 = 0$	$0\alpha_4 = 0$	$v^{31}v_{211}\alpha_4 = v^4 v_{31}$	$0\alpha_4 = 0$
v^{211}	0	$0\alpha_4 = 0$	$v^{211}\alpha_4 = v^4$	$0\alpha_4 = 0$	$0\alpha_4 = 0$	$0\alpha_4 = 0$	$0\alpha_4 = 0$	$0\alpha_4 = 0$
v^{22}	0	$0\alpha_4 = 0$	$0\alpha_4 = 0$	$v^{22}\alpha_4 = v^{22}$	$0\alpha_4 = 0$	$0\alpha_4 = 0$	$0\alpha_4 = 0$	$0\alpha_4 = 0$
v^{1111}	0	$0\alpha_4 = 0$	$0\alpha_4 = 0$	$0\alpha_4 = 0$	$v^{1111}\alpha_4 = v^{1111}$	$0\alpha_4 = 0$	$0\alpha_4 = 0$	$0\alpha_4 = 0$
$v^4 v_{31}$	0	$v^4 v_{31}\alpha_4 = v^{31}v_{211}$	$0\alpha_4 = 0$	$0\alpha_4 = 0$	$0\alpha_4 = 0$	$0\alpha_4 = 0$	$\gamma\alpha_4 = \gamma$	$0\alpha_4 = 0$
$v^{31}v_{211}$	0	$0\alpha_4 = 0$	$v^{31}v_{211}\alpha_4 = v^4 v_{31}$	$0\alpha_4 = 0$	$0\alpha_4 = 0$	$0\alpha_4 = 0$	$0\alpha_4 = 0$	$0\alpha_4 = 0$
$v^4 v_{211}$	0	$\alpha_4 = 0$	$v^4 v_{211}\alpha_4 = v^4 v_{211}$	$0\alpha_4 = 0$	$0\alpha_4 = 0$	$0\alpha_4 = 0$	$0\alpha_4 = 0$	$0\alpha_4 = 0$

Table 13.8: composition of the anti-automorphism α_4 of D_4, part 2

ad(ii): $A = (A^-)^-$ is valid, and hence part (ii) is a consequence of part (i).

ad(iii)-(vii): These parts are a consequence of parts (i) and (ii).

ad(ix): Based on part (vi) the set $\beta Aut(A)$ is contained in $Ant(A)$. Let $\alpha \in Ant(A)$. We use part (v) the statement $\beta^{-1} \in Ant(A)$ is true, and based on part (viii) we deduce $\alpha\beta^{-1} \in Aut(A)$. Thus, $Ant(A) = \beta Aut(A)$ is valid. By using this identity and part (v) we deduce $Ant(A) = \beta^{-1} Aut(A)$. By inverting this equation we get $Ant(A)^{-1} = Aut(A)^{-1}\beta$, and by applying part (v) we deduce $Ant(A) = Aut(A)\beta$.

ad(x): Based on parts (v) and (xi) the identity $Ant(A) = \beta^{-1} Aut(A) = Aut(A)\beta^{-1}$ is valid, and thus part (x) i proven.

ad(xi): This statement is a consequence of part (ix).$\diamond$

Theorem 32 *Let K be a field, $char(K) = 0$ and $n \in \mathbb{N}$. The following statements are valid:*

(i) $Ant(D_1) = Aut(D_1)$, $Aut(D_1) = 1$

(ii) $Ant(D_2) = Aut(D_2)$, $Aut(D_2) \cong Inn(D_2) \times C_2$

(iii) $Ant(D_3) = \alpha_3 Aut(D_3)$, $Aut(D_3) = Inn(D_3)$

(iv) $Ant(D_4) = \alpha_4 Aut(D_4)$, $Aut(D_4) \cong Inn(D_4) \times C_2$

(v) *For $n \geq 5$ the identity $Ant(D_n) = \emptyset$ is valid.*

Proof: The proof is a consequence of propositions 21, 22 and 23 and of theorem 4.22 in [3].$\diamond$

We remark that T. Bauer determines within his thesis [3] the group of automorphism and the Lie algebra of derivations for Solomons algebra: in both cases they are mainly the inner ones. Corresponding theorems for the Solomon-Tit algebra are not known to the author.

13.3 Dimensions of projective indecomposable left and right ideals of $K\Pi_n$

Definitions 16 If $n \in \mathbb{N}$ and $Q := \{Q_1, \cdots, Q_k\}$ is an unordered set partition of $\underline{n}$, then the type of Q – symbolized by $Typ(Q)$ – is the unique partition of n which is associated to the word $\mid Q_1 \mid \cdots \mid Q_k \mid$. The set of unordered set partitions of $\underline{n}$ is defined by $\Pi_n.\diamond$

Proposition 24 Let $n \in \mathbb{N}$, $p \vdash n$ and $M_p := \{Q \in \Pi_n \mid Typ(Q) = p\}$. The identities $\mid M_p \mid = \frac{n!}{\prod_{i \in \mathbb{N}} \mu_i(p)! i!^{\mu_i(p)}}$ and $\mid \Pi_n \mid = \sum_{q \vdash n} \mid M_q \mid$ are valid.

Proof: The symmetric group S_n acts on Π_n by $\{Q_1, \cdots, Q_k\}\alpha := \{Q_1\alpha, \cdots, Q_k\alpha\}$ for all $\alpha \in S_n$ and $\{Q_1, \cdots, Q_k\} \in \Pi_n$. It is straightforward to prove that the orbits of this S_n-action are exactly the sets $M_q, q \vdash n$. Thus, the second identity is valid because the orbits form a set partition of Π_n. Let $Q := \{Q_1, \cdots, Q_k\} \in M_p$, and let $p_1, \cdots, p_l, t_1, \cdots, t_l \in \mathbb{N}$ such that $Typ(Q) = p = p_1^{t_1} \cdots p_l^{t_l}$ is true. The stabilizer of Q under S_n is isomorphic to the direct product of the groups $S_{t_i} \times (S_{p_i})^{t_i}, i \in \underline{l}$. Thus, the first part is proven and the proof is finished.$\diamond$

Proposition 25 Let K be a field and $n \in \mathbb{N}$. The following statements are valid:

(i) For all $Q \in \Pi_n^{<}$ the identity $dim_K(e_Q K\Pi_n) \leq dim_K(e_{(123\cdots n)}K\Pi_n)$ is true.

(ii) $dim_K(e_{(123\cdots n)}K\Pi_n) = \sum_{p \vdash n}(\mid p \mid -1)! \frac{n!}{\prod_{i \in \mathbb{N}} \mu_i(p)! i!^{\mu_i(p)}}$

(iii) For all $Q \in \Pi_n^{<}$ the identity $dim_K(K\Pi_n e_Q) \leq dim_K(K\Pi_n e_{(1,2,3,\cdots,n)})$ is valid.

(iv) $dim_K(K\Pi_n e_{(1,2,3,\cdots,n)}) = n!$

Proof: ad(i): For all $P, Q \in \Pi_n^{<}$ let $c_{P,Q} := dim(e_P K\Pi_n e_Q)$. Based on remarks 6.5 in [18] we have to prove for all $P, Q \in \Pi_n^{<}$ with $Q \leq P$ that $c_{P,Q} \leq c_{(123\cdots n),Q}$ is valid. Let $P := (P_1, \cdots, P_l), Q := (Q_1, \cdots, Q_k) \in \Pi_n^{<}$ with $Q \leq P$. We use theorem 6.4 in [18] to deduce that $c_{P,Q} = \prod_{j=1}^{l}(m_j - 1)!$ with $m_j := \mid \{i \mid i \in \underline{k}, Q_i \subseteq P_j\} \mid$ is valid. In particular $\sum_{j=1}^{l} m_j = k$ and

$c_{(123\cdots n),Q} = (l(Q)-1)!$ are true. We use the multinomial coefficients to derive:

$$
\begin{aligned}
c_{P,Q} &= \prod_{j=1}^{l}(m_j-1)! \\
&\leq \left(\sum_{j=1}^{l}m_j-1\right)! \\
&= (k-l)! \\
&\leq (k-1)! \\
&= (l(Q)-1)! \\
&= c_{(123\cdots n),Q}.
\end{aligned}
$$

ad(ii): The dimension of the right ideals $e_{(123\cdots n)}K\Pi_n$ is – based on remarks 6.5 in [18] – exactly $\sum_{Q\in\Pi_n^{\leq}}(l(Q)-1)!$. We use remark 7 to deduce that the set $\Pi_n^{\leq}$ can be identified with Π_n. Hence, based on proposition 24 we calculate:

$$
\begin{aligned}
dim_K(e_{(123\cdots n)}K\Pi_n) &= \sum_{Q\in\Pi_n^{\leq}}(l(Q)-1)! \\
&= \sum_{p\vdash n}\sum_{Q\in M_p}(l(Q)-1)! \\
&= \sum_{p\vdash n}\sum_{Q\in M_p}(\mid p\mid-1)! \\
&= \sum_{p\vdash n}\mid M_p\mid(\mid p\mid-1)! \\
&= \sum_{p\vdash n}(\mid p\mid-1)!\frac{n!}{\prod_{i\in\mathbb{N}}\mu_i(p)!i!^{\mu_i(p)}}.
\end{aligned}
$$

ad(iii) + (iv): Let $Q \in \Pi_n^{\leq}$. The dimension of the left ideals $K\Pi_n e_Q$ is – based on theorem 5.4 in [18] – exactly $l(Q)!$. Thus, parts (iii) and (iv) are proven.$\diamond$

Remark 21 For $n \in \underline{5}$ we lists the dimensions $r_n := dim_K(e_{(123\cdots n)}K\Pi_n)$ and $l_n := dim_K(K\Pi_n e_{(1,2,3,\cdots,n)})$ within table 13.9. These dimensions are calculated based on proposition 25. The fraction $\frac{r_n}{l_n}$ can be calculated based on proposition 25, and it is $\sum_{p\vdash n}\frac{(\mid p\mid-1)!}{\prod_{i\in\mathbb{N}}\mu_i(p)!i!^{\mu_i(p)}}$. For the special partitions $i1^{n-i}, i =$

n	r_n	l_n
1	1	1
2	2	2
3	6	6
4	26	24
5	150	120.

Table 13.9: dimensions of projective indecomposable right and left ideals of Π_n

$2, \cdots, n$ of n we calculate the fractions $b_p := \dfrac{(|p|-1)!}{\prod\limits_{i\in\mathbb{N}} \mu_i(p)! i!^{\mu_i(p)}}$ and their sum. The limit is $e - 2^4$ (see table 13.10).

13.4 Anti-automorphism of $K\Pi_n$

Definition 2 Let M be a monoid. We call $Ant(M)$ the set of anti-automorphism of M.◇

Proposition 26 *Let K be a field and $n \in \mathbb{N}_{\geq 4}$. $K\Pi_n$ and Π_n do not possess any anti-automorphism: $Ant(K\Pi_n) = \emptyset = Ant(\Pi_n)$.*

<u>Proof:</u> Let $r_n := dim_K(e_{(123\cdots n)}K\Pi_n)$ and $l_n := dim_K(K\Pi_n e_{(1,2,3,\cdots,n)})$. Based on proposition 25 the value r_n resp. l_n is the maximal dimension of projective indecomposable right resp. left ideals. If an anti-automorphism of $K\Pi_n$ would exists, then $r_n = l_n$ would be valid. We use remark 21 to assume $n \geq 6$. Let us focus on the partitions $21n - 2, 31^{n-3}$ and 41^{n-4}. We get – by using remark 21 and proposition 25 – that $\frac{r_n}{l_n} \geq \frac{17}{24}$ is true. $\frac{17}{24} > \frac{7}{10}$ is valid, and if we add the partition 221^{n-4}, then we get $\frac{r_n}{l_n} > \frac{7}{10} + \frac{n-3}{8}$. For $n \geq 6$ is the inequality $\frac{n-3}{8} > \frac{3}{10}$ valid, and thus we have proven $r_n > l_n$. This contradiction finishes the proof for $K\Pi_n$. Any anti-automorphism of Π_n would extend linear to one of $K\Pi_n$.◇

The rest of this section is dedicated to the analysis for small values of n with respect to the existence of anti-automorphism of $K\Pi_n$.

---

[4]Leonhard Euler

$p \vdash n$	b_p	$\sum\limits_{p \vdash n} b_p$
21^{n-2}	$\frac{1}{2}$	$\frac{1}{2}$
31^{n-3}	$\frac{1}{6}$	$\frac{4}{6}$
41^{n-24}	$\frac{1}{24}$	$\frac{17}{24}$
51^{n-2}	$\frac{1}{120}$	$\frac{86}{120}$
61^{n-2}	$\frac{1}{720}$	$\frac{517}{720}$
$\vdots$		
$(n-1)1$	$\frac{1}{(n-1)!}$	$\sum\limits_{k=2}^{n-1} \frac{1}{k!}$
n	$\frac{1}{n!}$	$\sum\limits_{k=2}^{n} \frac{1}{k!} \;\to\; e-2.$

For the partition 221^{n-4} the identity $b_p = \frac{n-3}{8}$ is valid.$\diamond$

Table 13.10: approximation of $\frac{r_n}{l_n}$ within $K\Pi_n$

$\wedge$	1	e_2	e_3
1	1	e_2	e_3
e_2	e_2	e_2	e_2
e_3	e_3	e_3	$e_3,$

Table 13.11: composition on Π_2

Proposition 27 *Let K be a field. The following statements are valid:*

(i) $Ant(K\Pi_1) = Aut(K\Pi_1) = 1$

(ii) $Ant(\Pi_1) = Aut(\Pi_1) = 1$

(iii) *The linearization β_2 defined by $1\beta_2 = 1$, $(2,1)\beta_2 = 1-(2,1)$ and $((2,1)-(1,2))\beta_2 = (2,1) - (1,2)$ is an anti-automorphism of $K\Pi_2$.*

(iv) Π_2 *possesses no anti-automorphism.*

(v) *The automorphism group of $K\Pi_2$ is exactly the set of inner automorphism with respect to the subgroup $1 + rad(K\Pi_2)$.*

(vi) *If M, N are monoids possessing isomorphic monoid algebras KM and KN, then M and N must not be isomorphic.*

(vii) $\{1, 1-e_2, 1-2e_2+e_3\}$ *and Π_2 are anti-isomorphic submonoids of $K\Pi_2$.*

(viii) Π_3 *and $K\Pi_3$ possess no anti-automorphism.*

Proof: ad(i)+(ii): Both parts are a consequence of the fact $\mid \Pi_1 \mid = 1$.

ad(iii)+(iv): Let $e_2 := (2,1)$ and $e_3 := (1,2)$. The sets $\{1, e_2, e_3\}$ and $\{1 - e_2, e_2, e_2 - e_3\}$ are K-bases of $K\Pi_2$. Part (iii) is deductable by the following calculation: The 'transpose table' of table 13.12 is exactly table 13.13. Within the second and third table the 1-element is not included, in the first table 13.11 the composition on Π_2 is presented.

Now we prove that Π_2 possesses no anti-automorphism. We assume that an anti-automorphism α of Π_2 exists. The identity $1\alpha = 1$ is valid. Π_2 is non-commutative ($e_2e_3 = e_2 \neq e_3 = e_3e_2$), and thus $\alpha \neq id_{\Pi_2}$ must be valid. Hence, $e_2\alpha = e_3$ and $e_3\alpha = e_2$ is true. The calculation $(e_2e_3)\alpha = e_2\alpha = e_3 \neq e_2 = e_2e_3 = (e_3\alpha)(e_2\alpha)$ is a contradiction.

$(ab)\beta_2$	e_2	$e_2 - e_3$
e_2	$1e_2\beta_2 = e_2$	$0\beta_2 = 0$
$e_2 - e_3$	$(e_2 - e_3)\beta_2 = e_2 - e_3$	$0\beta_2 = 0,$

Table 13.12: anti-automorphism β_2 on $K\Pi_2$, part 1

$(b\beta_2)(a\beta_2)$	$1 - e_2$	$e_2 - e_3$
$1 - e_2$	$1 - e_2$	$e_2 - e_3$
$e_2 - e_3$	0	$0.$

Table 13.13: anti-automorphism β_2 on $K\Pi_2$, part 2

ad(v): Based on theorem 3 the nilradical of $K\Pi_2$ is spanned by the element $r := e_{(1,2)} - e_{(2,1)}$, and the elements $e := e_{(12)} = (12) - (1,2)$ and $f := e_{(1,2)} = (1,2)$ span a radical complement. Let α be an automorphism of $K\Pi_2$. Theorem 29 is used to derive the existence of elements $k, l, m \in K \setminus \{0\}$ and $x, y \in 1 + rad(K\Pi_2)$ such that $r\alpha = kr$, $e\alpha = le^x$ and $f\alpha = mf^y$ are valid. In addition, let $p, q \in K$ such that $x = 1 + pr$ and $y = 1 + qr$ is true. Within table 13.14 the composition of the element $e_{(12)}, e_{(1,2)}$ and $e_{(2,1)}$ is presented. We derive the following tables 13.15 and 13.16 for the automorphism α. We deduce $k = l = m = 1$ and $p = q$. In particular, $x = y$ is valid. Because of $r^2 = 0$ we get $r = r^x$, and thus α is the conjugation with $1 + pr$.

ad(vi): Based on (iii) and (iv) the algebras $K\Pi_2$ and $(K\Pi_2)^-$ but not the monoids Π_2 and $(\Pi_2)^-$ are isomorphic.

ad(vii): The set within part (vii) is the image of Π_2 under β_2. We use (iii) and (iv) to finish the proof of (vii).

ad(viii): We assume that $K\Pi_3$ possesses an anti-automorphism α. Let $LH1(K\Pi_3)$ resp. $RH1(K\Pi_3)$ the set of those elements x of $K\Pi_3$ such that

$\wedge$	$e_{(12)}$	$e_{(1,2)}$	$e_{(2,1)}$
$e_{(12)}$	$e_{(12)}$	0	$e_{(2,1)} - e_{(1,2)}$
$e_{(1,2)}$	0	$e_{(1,2)}$	$e_{(1,2)}$
$e_{(2,1)}$	0	$e_{(2,1)}$	$e_{(2,1)}.$

Table 13.14: composition for a basis of $K\Pi_2$

$K\Pi_3 \cdot x = \langle x \rangle_K$ resp. $x \cdot K\Pi_3 = \langle x \rangle_K$ is valid (the generator of the one-dimensional principal left resp. right ideals of $K\Pi_3$). α is a bijection between these two. If one of the sets is a K-subspace, then the other one is a K-space, too, and their K-dimension is identical. We prove that $RH1(K\Pi_3)$ is 6-dimensional and $LH1(K\Pi_3)$ is of dimension at most 5. The necessary calculations can be done by using the composition table presented within chapter 1 of the 13-dimensional algebra $K\Pi_3$. The reader may do this within exercise 179. The concept is to present an element x of $RH1(K\Pi_3)$ resp. of $LH1(K\Pi_3)$ by the 13-dimensional basis $\{e_1, \ldots, e_{13}\}$ and use the condition $x \cdot K\Pi_3 = \langle x \rangle_K$ resp. $K\Pi_3 \cdot x = \langle x \rangle_K$ and again the basis. The consequence is a system of linear equation which can be solved as desired.$\diamond$

Theorem 33 *Let K be a field and $n \in \mathbb{N}$. The following statements are valid:*

(i) $Ant(K\Pi_1) = Aut(K\Pi_1) = 1$

(ii) $Ant(\Pi_1) = Aut(\Pi_1) = 1$

(iii) $Ant(K\Pi_2) = \beta_2 Aut(K\Pi_2)$, $Aut(K\Pi_2) = Inn(K\Pi_2)$

(iv) For $n \geq 3$ the identity $Ant(K\Pi_n) = \emptyset$ is true.

(v) For $n \geq 2$ the statement $Ant(\Pi_n) = \emptyset$ is valid.

Proof: The statements are a direct consequence of propositions 26, 27 and 23.$\diamond$

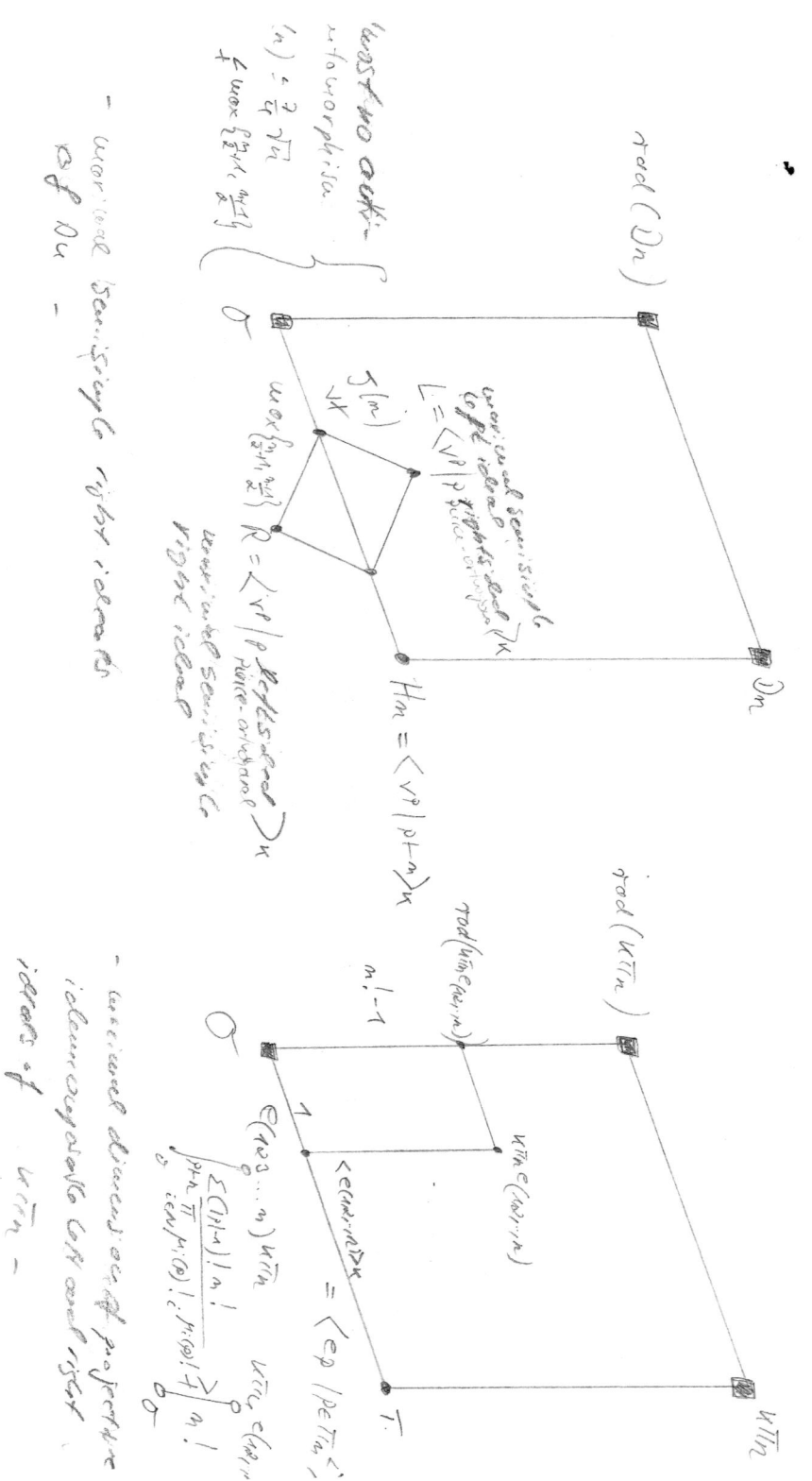

Chapter 13

rad($\mathcal{O}_n$)

$\mathcal{O}_n$

rad($u\Pi_n$)

$u\Pi_n$

rad($u\Pi_n(u_1...u_n)$)

$u\Pi_n \in (u_1...u_n)$

$\langle e_{(u_1...u_n)} \mathbb{Z}_K$

$= \langle e_p | p \in \Pi_n \subset$

maximal semisimple
top ideal

$L = \langle V^p | p \text{ rights ideal} \rangle_K$
since orientation

$H_n = \langle V^p | p + \lambda \rangle_K$
$\eta, i-1$

$J_{(m)}$
$\forall K$

$\max \{\frac{2m}{2}, m+j\} R = \langle V^p | p \text{ 2-sided ideal} \rangle_K$
more orientation

maximal semisimple
right ideal

"cosemisimple
automorphism

$\ln) = \frac{2}{4} \frac{\pi}{\pi}$
$\neq \max \{\frac{2m}{2}, \frac{m+j}{2}\}$

maximal semisimple
right ideals

of $\mathcal{O}_4$ —

maximal dimensional left projective
ideausalgebras left and right

ideals of $u\Pi_n$ —

$\frac{\sum_{(1,2,...,n)} u\Pi_n}{p_{\mu_i} \prod_i i \cdot p_{i(q)} \cdot \vec{z}} \} \eta . !$

$\sum_{(1,2,...,n)} u\Pi_n$

$u\Pi_n \in (u_1...u_n)$

$1, -1$

1

T_*

$(ab)\alpha$	r	e	f
r	0	0	kr
e	kr	le^x	0
f	0	0	$mf^y,$

Table 13.15: composition for automorphism α of $K\Pi_2$, part 1

$(a\alpha)(b\alpha)$	kr	le^x	mf^y
kr	0	0	mk^2r
le^x	klr	l^2e^x	$lm(p-q)r$
mf^y	0	0	$m^2f^y.$

Table 13.16: composition for automorphism α of $K\Pi_2$, part 2

13.5 Open-ended questions

- What are the anti-automorphism of a monoid algebra for a finite idempotent monoid?

- What are the anti-automorphism of a finite idempotent monoid?

- Let $n \in \mathbb{N}$. What are the automorphism of Π_n?

- Let K be a field and $n \in \mathbb{N}$. What are the automorphism of $K\Pi_n$?

- Is the reflection s_n an inner or outer automorphism of $K\Pi_n$?

- Let M be a finite idempotent monoid. What are the automorphism of M?

- Let K be a field and M a finite idempotent monoid. What are the automorphism of KM?

- Let K be a field and $n \in \mathbb{N}$. What are the derivations of $K\Pi_n$?

- Let K be a field and M a finite idempotent monoid. What are the derivations of KM?

- Let K be a field, $n \in \mathbb{N}$ and $P, Q \in \Pi_n$. If $P < Q$ is valid, then $dim_K(e_P K\Pi_n) \leq dim_K(e_Q K\Pi_n)$ is true or not? What is the answer for the corresponding left ideals? What are the answers if e_P and e_Q are replaced by P and Q?

13.6   Exercises

Exercise 179 *Execute the calculations of part (viii) of proposition 27 within the 13-dimensional algebra $K\Pi_3$!*

Exercise 180 *For all $n \in \mathbb{N}$ the identity $B(n) = \mid \Pi_n \mid$ is valid.*

Exercise 181 *Let A be a K-algebra. Are $(A^-)^-$ and A isomorphic or identical?*

Exercise 182 *Does the fraction $\frac{r_n}{l_n}$ within remark 21 converge? What about the difference $r_n - l_n$? Is it possible to determine r_n and l_n explicitly?*

Exercise 183 *Let K be a field and $n \leq 3$. Is the reflection s_n and inner or outer automorphism of $K\Pi_n$?*

Exercise 184 *Let K a field and G a finite group. Prove that KG and $(KG)^-$ are isomorphic. Is this result also true for an arbitrary finite semigroup G?*

Exercise 185 *Prove the following statement: If $n = \prod\limits_{i=1}^{s} p_i^{r_i}$ is the decomposition into prime numbers of n, then the existence and uniqueness of the decompositions yields to $\tau(n) = \prod\limits_{i=1}^{r_i}(r_i + 1)$.*

Exercise 186 *Let K be a field and $n \in \mathbb{N}$. What is the maximal dimension of the semisimple resp. simple left, right and two-sided ideals of $K\Pi_n$?*

Exercise 187 *Let K be a field and $n \in \mathbb{N}$. $K^{n \times n}$ and its inverse algebra are isomorphic! Is this statement true if we replace $K^{n \times n}$ by $A^{n \times n}$ such that A is a finite-dimensional associative unitary K-algebra?*

Exercise 188 *For a field K calculate the dimensions of maximal semisimple resp. separable resp. simple right ideals of $K\Pi_{17}$!*

Exercise 189 *For a field K calculate the dimensions of maximal semisimple resp. separable resp. simple left ideals of $K\Pi_{17}$!*

Exercise 190 *For a field K calculate the dimensions of maximal semisimple resp. separable resp. simple ideals of $K\Pi_{17}$!*

Exercise 191 *For a field K of characteristic 0 calculate the dimensions of maximal semisimple resp. separable resp. simple right ideals of D_{17}!*

Exercise 192 *For a field K of characteristic 0 calculate the dimensions of maximal semisimple resp. separable resp. simple left ideals of D_{17}!*

Exercise 193 *For a field K of characteristic 0 calculate the dimensions of maximal semisimple resp. separable resp. simple ideals of D_{17}!*

Exercise 194 *An associative unitary K-algebra possesses 42 automorphism and at least on anti-automorphism. Count the number of anti-automorphism!*

Exercise 195 *Does an arbitrary associative unitary K-algebra possess an anti-automorphism?*

Exercise 196 *True or false: If an associative unitary K-algebra possesses an antiautomorphism, then one of order 2 exists.*

Exercise 197 *Let K be a field. Determine the automorphism of $K\Pi_1$, $K\Pi_2$ and $K\Pi_3$! Are the reflections s_1, s_2 and s_3 inner or outer automorphism?*

Exercise 198 *Let K be a field. Determine all derivations of $K\Pi_1$, $K\Pi_2$ and $K\Pi_3$.[5]*

Exercise 199 *Let K be a field and $n \in \mathbb{N}$. For $n = 1, ..., 20$ decide whether $K\Pi_n$ and its inverse algebra are isomorphic!*

Exercise 200 *Let $n \in \mathbb{N}$. For $n = 1, ..., 20$ decide whether Π_n and its inverse monoid are isomorphic!*

Exercise 201 *Let K be a field of characteristic 0 and $n \in \mathbb{N}$. For $n = 1, ..., 20$ decide whether D_n and its inverse algebra are isomorphic!*

Exercise 202 *Determine all $n \in \mathbb{N}$ such that $\frac{7}{4}\sqrt[2]{n} < \frac{n+1}{2} \le \frac{n}{2} + 1$ is valid.*

Exercise 203 *Let K be a field and $n \le 3$. Does the algebra of lower triangular matrices of $K^{n \times n}$ possess an anti-automorphism (of order 2)? (Tip: see [22]!) Is it possible to describe all anti-automorphism, automorphism and derivations? Do you have a conjecture for an arbitrary n?*

---

[5]The derivations of Solomons algebra over a field of characteristic 0 are determined by Thorsten Bauer in [3].

Chapter 14

Values of irreducible characters of $K\Pi_n$

Within this chapter we focus on the following topics related to irreducible characters of monoid algebras related to finite idempotent monoids:

- determination of irreducible characters of $K\Pi_n$ by Kenneth Brown

- description of values 0 and 1 of the irreducible characters based on sub-semigroups

- determination of the number of elements of these sub-semigroups for Π_n

- consequences for the right ideals $P \cdot K\Pi_n$ and $e_P \cdot K\Pi_n$

- consequences for the left ideals $K\Pi_n \cdot P$ and $K\Pi_n \cdot e_P$.

14.1   Sub-semigroups and irreducible characters of $K\Pi_n$

Definitions 17 Let H be a semigroup and $x, y \in H$. We define:

- $Hx := \{zx \mid z \in H\}$ - left-multiples of x

- $xH := \{xz \mid z \in H\}$ - right-multiples of x

- $H_{\leq y} := \{z \mid z \leq y\}$ - set of $\leq$-elements of y

- $H_{\geq y} := \{z \mid y \geq z\}$ - set of $\geq$-elements of y

- $H_{>y} := H_{\geq y} \setminus [y]_\sim$ - set of $\leq$-elements of but not associated elements to y

- $H_{<y} := H_{\leq y} \setminus [y]_\sim$ - set of $\geq$-elements of but not associated elements to y

- $H_{>y<} := H \setminus (H_{\leq y} \cup H_{\geq y})$ - set of non-comparable elements to y.$\diamond$

Remark 22 *Let* $n \in \mathbb{N}$ *and* $P, Q \in \Pi_n$. $P \leq Q$ *is valid if and only if* $P = P \wedge Q$ *is true, because:* $P \wedge Q \wedge P = P \wedge Q$.$\diamond$

Lemma 10 *(Kenneth Brown, [7]) Let* M *be an idempotent monoid. The following statements are valid:*

(i) $\sim$ *is an equivalence relation on* M.

(ii) $m \leq 1$ *for all* $m \in M$

(iii) $\leq$ *is reflexive and transitive on* M.

(iv) For all $x, y, z \in M$ *the identity* $xy \leq z$ *is valid if and only if the statements* $x \leq z$ *and* $x \leq y$ *are true.*$\diamond$

Remark 23 Let H be a semigroup and $\leq$ a relation on H fulfilling the statements of parts (iii) and (iv) of lemma 10. We define the inverse-relation $a \geq b$ by $b \leq a$ on H. This relation fulfills the same statements of the mentioned lemmas.$\diamond$

Lemma 10 and remark 23 yield to the following corollary:

Corollary 21 *Let* M *be an idempotent monoid. The following statements are valid:*

(i) $1 \geq m$ *for all* $m \in M$

(ii) $\geq$ *is reflexive and transitive on* M.

(iii) For all $x, y, z \in M$ *the identity* $xy \geq z$ *is valid if and only if the statements* $x \geq z$ *and* $x \geq y$ *are true.*$\diamond$

A direct consequence of lemma 10 and remark 23 is the following:

Corollary 22 *Let* M *be an idempotent monoid and* $a \in M$. *The following statements are valid:*

(i) $M_{\leq a}$ *is a sub-semigroup of* M.

(ii) $M_{\geq a}$ *is a sub-monoid of M.*

(iii) $[a]_\sim$ *is a sub-semigroup of M.*

(iv) $[a]_\sim = M_{\leq a} \cap M_{\geq a}$

(v) $M_{>a}$ *is a sub-semigroup of M.*

(vi) $M_{<a}$ *is a sub-semigroup of M.*

(vii) $M_{>a<}$ *is a sub-semigroup of M.*

(viii) M *is a disjoint union of the sub-monoids $[a]_\sim, M_{>a}, M_{<a}$ and $M_{>a<}$.* ⋄

Definition 3 Let K be a field, M a finite idempotent monoid and $x \in M$. We define

$$\chi_x : M \longrightarrow K, m \mapsto \begin{cases} 1 & : & m \in M_{\geq x} \\ 0 & : & m \notin M_{\geq x} \end{cases}.$$

In addition, let $\hat{\chi}_x$ be the linearization of χ_x on KM. ⋄

Remark 24 Let K be a field, M a finite idempotent monoid and $x \in M$. $1\hat{\chi}_x = 1$ is valid, and the kernel of $\hat{\chi}_x$ is of codimension 1. Let $z := \sum_{a \in M} k_a a \in KM$. $z\hat{\chi}_x = 0$ is valid if and only if $\sum_{a \in M_{\geq x}} ka = 0$ is true. Thus, $Kern\hat{\chi}_x = Aug(KM_{\geq x}) \oplus KM_{<x} \oplus KM_{>x<}$ is valid. ⋄

The following theorem is valid:

Theorem 34 *(Kenneth Brown, [7]) Let K be a field and M a finite idempotent monoid. The following statements are valid:*

(i) For all $x \in M$ the function $\hat{\chi}_x$ is an epimorphism from KM on K.

(ii) For all $x, y \in M$ the identity $\hat{\chi}_x = \hat{\chi}_y$ is valid if and only if $x \sim y$ is true.

(iii) If R is a complete set of representatives for the equivalence classes of $\sim$, then $\{\hat{\chi}_x \mid x \in R\}$ is the set of irreducible characters of KM. ⋄

Definition 4 Let M be an idempotent monoid. An element m is maximal resp. minimal with respect to $\leq$ if and only if for all $x \in M$ the identity $x \leq m$ resp. $m \leq x$ is valid. ⋄

Remark 25 Let M be an idempotent monoid. If m is a maximal resp. minimal element with respect to $\leq$, then $[m]_\sim$ is the of all maximal resp. minimal element of M because all minimal resp. maximal elements are associated (by definition).

Let $n \in \mathbb{N}$. Exactly one maximal element exists which is 1_{Π_n}. In addition, exactly $n!$ minimal elements exist which are the associated elements of $(1, 2, 3, ..., n)$ in Π_n.⋄

Example 10 *Let K be a field and $n \in \mathbb{N}$. Because of remark 25 the following statements are valid:*

(i) $(\Pi_n)_{\geq 1} = \{1\}$, $(\Pi_n)_{<1} = M \setminus \{1\}$ and $(\Pi_n)_{>1<} = \emptyset$

(ii) $Kern\hat{\chi}_1 = \langle m \mid m \in \Pi_n, m \neq 1 \rangle_K$

(iii) $(\Pi_n)_{\geq(1,2,...,n)} = M$, $(\Pi_n)_{<(1,2,...,n)} = \emptyset$ and $(\Pi_n)_{>(1,2,...,n)<} = \emptyset$

(iv) $Kern\chi_{(1,\hat{2},...,n)} = \langle m - 1 \mid m \in \Pi_n \rangle_K$.⋄

Definition 5 Let $\sim$ be an equivalence relation on a set R. If T is a finite subset of R which is a disjoint union of equivalence classes of $\sim$, then we denote by $T_\sim$ the set of these classes and $\mid T_\sim \mid$ their number.⋄

Remark 26 Let M be an idempotent monoid and $x \in M$. Based on lemma 10 and corollary 21 the sets $M_{>y<}, M_<, M_{>x}, M_{\leq x}, M_{\geq x}$ are disjoint unions of equivalence classes of $\sim$.⋄

Example 11 We focus on the irreducible characters of $K\Pi_3$, and the corresponding values of the previous remark are presented within the next table. The elements of $(\Pi_3)_{\geq P}$ are exactly the elements of Π_3 such that χ_P is of value 1. Likewise, the elements of Π_3 send by χ_P to 0 are the elements within the disjoint sets $(\Pi_3)_{<P}$ and $(\Pi_3)_{>P<}$. Π_3 is the union of these three sets. For entering the values in the mentioned table we need to know the number of equivalence classes within these three sets. The number of elements within one equivalence class is the factorial of the length of one (each) element contained in the class.⋄

Inspired by this example we want to determine how many equivalence classes and elements are $\leq, <$ or $\geq, >$ than an element $P \in \Pi_n$. The non-comparable ones are the differences of the sets Π_n, $(\Pi_n)_{\geq P}$ and $(\Pi_n)_{<P}$. The sets $(\Pi_n)_{<P}$ and $(\Pi_n)_{\leq P}$ differ only by the class $[P]_\sim$. Manfred Schocker has proven in [18], Remarks 6.4 the following insights about the set $(\Pi_n)_{\leq P}$:

$\chi_P, P \in \Pi_3^<$	$\|(\Pi_3)_{\geq P}\|$	$\|((\Pi_3)_{\geq P})_\sim\|$	$\|(\Pi_3)_{<P}\|$	$\|((\Pi_3)_{<P})_\sim\|$	$\|(\Pi_3)_{>P<}\|$	$\|((\Pi_3)_{>P>})_\sim\|$
$(1,2,3)$	13	5	0	0	0	0
$(12,3)$	2	3	6	1	4	2
$(1,23)$	2	3	6	1	4	2
$(2,13)$	2	3	6	1	4	2
(123)	1	1	12	4	0	0

Table 14.1: values of characters within $K\Pi_3$

Proposition 28 *Let K be a field, $n \in \mathbb{N}$ and $P = (P_1, \cdots, P_l) \in \Pi_n$. The function*

$$\Pi_{l(P)} \longrightarrow (\Pi_n)_{\geq P}, (I_1, \cdots, I_l) \mapsto (\bigcup_{i_1 \in I_1} P_{i_1}, \cdots, \bigcup_{i_l \in I_l} P_{i_l})$$

is a monoid-isomorphism which is compatible with $\sim$. In particular, $\mid (\Pi_n)_{\geq P} \mid = \mid \Pi_{l(P)} \mid$ and $\mid ((\Pi_n)_{\geq P})_\sim \mid = \mid (\Pi_{l(P)})_\sim \mid$ are valid. The annihilator of $K\Pi_n \cdot e_P$ is a complement of $K(\Pi_n)_{\geq P}$ in $K\Pi_n.\diamond$

Thus, the determination of the number of equivalence classes and elements related to $(\Pi_n)_{\geq P}$ is reduced to the determination within $\Pi_{l(P)}$. These numbers are already presented in the second chapter about dimensions. We illustrate the result by an example:

Example 12 Let $n = 10$ and $P := (123, 45, 6, 9(10), 78)$. The length of P is exactly 5. We use remark 8 and example 2 to derive that exactly 541 elements $\geq$ as P exist. The set of these elements are a disjoint union of 52 equivalence classes with respect to $\sim.\diamond$

Proposition 29 *Let $n \in \mathbb{N}$, $P = (P_1, \cdots, P_r) \in \Pi_n$ and $Typ(P) = (p_1, \cdots, p_r)$. The following statements are valid:*

(i) $(\Pi_n)_{\leq p}$ consists of $\prod_{i=1}^{r} \mid \Pi_{p_i}^< \mid = \prod_{i=1}^{r} B(p_i)$ equivalence classes.

(ii) The number of elements within $(\Pi_n)_{\leq p}$ is countable by the formula

$$\sum_{X_1 \in \Pi_{P_1}^<} \cdots \sum_{X_r \in \Pi_{P_r}^<} (l(X_1) + \cdots l(X_r))!$$
$$= \sum_{i_1=1}^{p_1} \cdots \sum_{i_r=1}^{p_r} S(p_1, i_1) \cdots S(p_r, i_r)(i_1 + \cdots i_r)!.$$

Proof: $Q = (Q_1, ..., Q_r) \leq P = (P_1, ..., P_l)$ is valid if and only if $Q = QP$ is true (Q is a refinement of P.). This is equivalent to the fact that for all Q_i an element P_j exists such that $Q_i \subseteq P_j$ is valid. On the direct product $\Pi_{P_1} \times \cdots \times \Pi_{P_l}$ we define $\sim$ componentwise. In addition, $\sim$ is also defined on $\Pi_{\leq P}$ because this sub-monoid is a disjoint union of $\sim$-classes. We focus on the function $\varphi : \Pi_{P_1} \times \cdots \times \Pi_{P_l} \longrightarrow \Pi_{\leq P}$ that is the concatenation of partitions. φ is an injective monoid-homomorphism. (In general, it is not surjective: see exercises.). We prove that it is a bijection on the $\sim$-classes, and this statement finishes the proof of this proposition. If Q and R are associated within the direct product, then they are associated within $\Pi_{\leq P}$, too: the components of Π_{P_j} can be transformed likewise by sorting. If $Q\varphi$ and $R\varphi$ are associated in $\Pi_{\leq P}$, then Q, R are – by definition of φ – associated within the direct product: they can be transformed into each other by re-arranging, but by using the disjointness of $P_1, ..., P_r$ all factors are associated. If $Q \in \Pi_{\leq P}$ is valid, then we need to define an associated element to Q which possesses a pre-image within the direct product. For this, re-arrange Q such that P_1 is at first, then $P_2,...$, and at the end P_l. This is possible based on the definition of $\Pi_{\leq P}$. Such an element is a pre-image in the direct product.$\diamond$

Example 13 Let K be a field and $P := (12, 348, 567) \in \Pi_8$. We determine the number of elements and equivalence classes within $(\Pi_8)_{\leq P}$. For $P_1 := \{1, 2\}$, $P_2 := \{3, 4, 8\}$ and $P_3 := \{5, 6, 7\}$ the rule $P = (P_1, P_2, P_3)$ is valid. Π_{P_1} is isomorphic to Π_2, Π_{P_2} is isomorphic to Π_3 and Π_{P_3} is isomorphic to Π_3. Π_2 possesses 2 and Π_3 exactly 3 classes, and thus $(\Pi_8)_{\leq P}$ consist of $2 \cdot 5 \cdot 5 = 50$ classes.

Let $Q := (12, 345) \in \Pi_5$. $\Pi_{\{1,2\}}$ contains one element of length 1 and one element of length 2, $\Pi_{\{3,4,5\}}$ one element of length 1, three elements of length 2 and one element of length 3 within the system of representants $\Pi^{<}_{\{1,2\}}$ and $\Pi^{<}_{\{3,4,5\}}$. Thus, in $(\Pi_5)_{\leq Q}$ are $2! + 3! \cdot 3 + 4! + 3! + 3 \cdot 4! + 5! = 242$ elements situated in $1 \cdot 3 = 3$ equivalence classes.$\diamond$

Now we turn our focus on the principal one- and two-sided ideals generated by basis elements of Π_n resp. of $e_P, P \in \Pi_n$.

Theorem 35 *Let K be a field, $n \in \mathbb{N}$ and $P \in \Pi_n$. The following statements are valid:*

(i) $P \cdot K\Pi_n \subseteq K\Pi_n \cdot P$

(ii) In general, $P \cdot K\Pi_n$ is contained proper in $K\Pi_n \cdot P$.

(iii) $K\Pi_n \cdot P = K\Pi_n \cdot P \cdot K\Pi_n$ is an ideal.

(iv) $K\Pi_n \cdot P = K(\Pi_n)_{\leq P}$ is a semigroup-algebra related to a finite idempotent semigroup of K-dimension $| (\Pi_n)_{\leq P} |$, and $K\Pi_n \cdot (1 - P)$ is a left ideal complement of $K\Pi_n \cdot P$ in $K\Pi_n$.

(v) $\{PQ \mid Q \in \Pi_n\}$ is a sub-semigroup of $(\Pi_n)_{\leq P}$ which is also a monoid with 1-element P.

(vi) The left-multiplication λ_P with P on $K\Pi_n$ is an algebra-endomorphism with image $Bild(\lambda_P) = P \cdot K\Pi_n$.

(vii) $P \cdot K\Pi_n$ *is a monoid algebra related to the finite idempotent monoid* $\{PQ \mid Q \in \Pi_n\}$*, and the set* $(1 - P) \cdot K\Pi_n$ *is a right ideal complement of* $P \cdot K\Pi_n$ *in* $K\Pi_n$*.*

Proof: ad(i): For all $P, Q \in \Pi_n$ the identities $PQ \sim QP$, $PQ, QP \leq P$ and $PQ, QP \leq Q$ are valid. Thus, part (i) is a consequence of part (iv).

ad(ii): If $P := (1, 2, \cdots, n)$ is defined, then $P \cdot K\Pi_n$ is one-dimensional but $K\Pi_n \cdot P$ is $n!$-dimensional.

ad(iii): Based on part (i) we deuce that $K\Pi_n \cdot P$ is an ideal. Therefor. it is identical to the principal ideal $K\Pi_n \cdot P \cdot K\Pi_n$.

ad(iv): For all $Q \in \Pi_n$ the statement $QP \leq P$ is valid. If $Z \leq P$ is true, then $Z = ZP$ is valid. Thus, $\{QP \mid Q \in \Pi_n\} = (\Pi_n)_{\leq P}$ is true. P is an idempotent, and thus part (iv) is true.

ad(v): Let $Q, R \in \Pi_n$. $P(PQ) = PQ = (PQ)P$ and $(PQ)(PR) = (PQP)R = (PQ)R$ are valid.

ad(vi): see part (v)

ad(vii): see part (vi) and use the idempotency of $P\diamond$

Proposition 30 *Let K be a field, $n \in \mathbb{N}$ and $P, Q \in \Pi_n$. The following statements are valid:*

(i) $PQ \sim QP$

(ii) *Let $R, S \in \Pi_n$ such that $R \sim S$ is valid. The identities $K\Pi_n \cdot R = K\Pi_n \cdot S$ and $R \cdot K\Pi_n = S \cdot K\Pi_n$ are true.*

(iii) $K\Pi_n \cdot PQ = K\Pi_n \cdot QP$

(iv) $K\Pi_n PQ = (K\Pi_n \cdot P) \cap (K\Pi_n \cdot Q) = (K\Pi_n \cdot P)(K\Pi_n \cdot Q)$

(v) $PQ \cdot K\Pi_n = QP \cdot K\Pi_n$

(vi) $PQ \cdot K\Pi_n = (P \cdot K\Pi_n) \cap (Q \cdot K\Pi_n) = (P \cdot K\Pi_n)(Q \cdot K\Pi_n)$

(vii) $PQ \cdot K\Pi_n \subseteq K\Pi_n PQ$

Proof: The proof is based on a straightforward calculation and theorem 35.⋄

We use this proposition and theorem 35 to derive the following corollary by an induction argument:

Corollary 23 *Let K be a field, $n, r \in \mathbb{N}$, $\alpha \in S_r$ and $P_1, \cdots, P_r \in \Pi_n$. The following statements are valid:*

(i) $P_1 \cdots P_r \sim P_{1\alpha} \cdots P_{r\alpha}$

(ii) $K\Pi_n \cdot (P_1 \cdots P_r) = K\Pi_n \cdot (P_{1\alpha} \cdots P_{r\alpha})$

(iii) $K\Pi_n(P_{1\alpha} \cdots P_{r\alpha}) = \bigcap_{i=1}^{r}(K\Pi_n \cdot P_i) = \prod_{i=1}^{r}(K\Pi_n \cdot P_i)$

(iv) $(P_1 \cdots P_r) \cdot K\Pi_n = (P_{1\alpha} \cdots P_{r\alpha}) \cdot K\Pi_n$

(v) $(P_{1\alpha} \cdots P_{r\alpha}) \cdot K\Pi_n = \bigcap_{i=1}^{r}(P_i \cdot K\Pi_n) = \prod_{i=1}^{r}(P_i \cdot K\Pi_n)$

(vi) $(P_{1\alpha} \cdots P_{r\alpha}) \cdot K\Pi_n \subseteq K\Pi_n(P_{1\alpha} \cdots P_{r\alpha}).$⋄

Manfred Schocker has proven within [18], theorem 5.4 and remarks 6.5(1):

Theorem 36 *Let K be a field, $n \in \mathbb{N}$, $P := (P_1, \cdots, P_l) \in \Pi_n$ and $Typ(P) := (p_1, \cdots p_l)$. The following statements are valid:*

(i) $K\Pi_n \cdot e_P$ *possesses the basis* $\{e_Q \mid Q \sim P\}$ *and is of K-dimension $l(P)!$.*

(ii) $e_P \cdot K\Pi_n$ *possesses the K-dimension* $2^l \cdot \mid \Pi_{p_1 - 1} \mid \cdots \mid \Pi_{p_l - 1} \mid$.

(iii) $K\Pi_n \cdot e_P \cdot K\Pi_n$ *possesses the K-span* $\{e_{RQ} \mid P \sim Q \leq R\}.$⋄

Chapter 14.

18 elements
5 orders

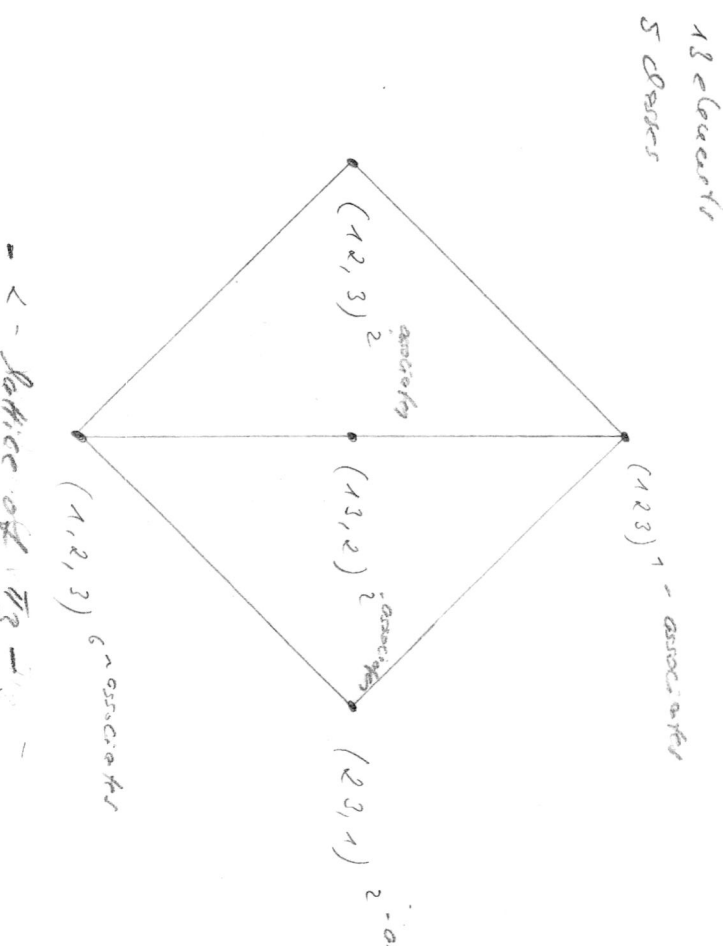

$(12, 3)^2$ associate

$(123)^1$ - associates

$(13, 2)^2$ - associate

$(1, 2, 3)^6$ - associates

$(23, 1)^2$ - associates

1 — Lattice of $\overline{F_3}$ —

14.2 Open-ended questions

- Let K be a field and M a finite idempotent monoid. Is it possible to generalize the results of this chapter to KM?

- What is the dimension of $K\Pi_n \cdot e_P \cdot K\Pi_n$?

14.3 Exercises

Exercise 204 *Let K be a field and $P := (3456, 17, 2) \in \Pi_7$. Determine the number of elements within $(\Pi_7)_{\leq P}$ and analyzed how many classes are contained in this set!*

Exercise 205 *Is the function within proposition (on the elements but not on the classes with respect to $\sim$) 29 surjective? On what terms is the function surjective? Is the function surjective within example 13 or within exercise 204?*

Exercise 206 *As presented within example 11 do the same for $n = 4$ and draw a $\leq$-diagram of Π_4!*

Exercise 207 *Let K be a field and $n \in \mathbb{N}$. For which $P \in \Pi_n$ the following statements are valid:*

(i) $P \cdot K\Pi_n \subseteq K\Pi_n \cdot P$?

(ii) $e_P \cdot K\Pi_n \subseteq K\Pi_n \cdot e_P$?

(iii) $P \cdot K\Pi_n = K\Pi_n \cdot P$?

(iv) $e_P \cdot K\Pi_n = K\Pi_n \cdot e_P$?

(v) $K\Pi_n \cdot P \subseteq P \cdot K\Pi_n$?

(vi) $K\Pi_n \cdot e_P \subseteq e_P \cdot K\Pi_n$?

Exercise 208 *Let K be a field, A an associative finite-dimensional K-algebra and B a basis of A. True or false: A is a Duo algebra if for all $b \in B$ the identity $bA \subseteq Ab$ is valid.*

Exercise 209 *Let K be a field, A an associative finite-dimensional K-algebra and B a basis of A. True or false: A is a Duo algebra if for all $b \in B$ the identity $Ab \subseteq bA$ is valid.*

Exercise 210 *Let K be a field, A an associative finite-dimensional K-algebra and B a basis of A. True or false: A is a Duo algebra if for all $b \in B$ the identity $bA = Ab$ is valid.*

Exercise 211 *For $n = 9$ and $P := (3467, 12, 569, 8)$ determine all quantities as done within example 12! What is the dimension of the annihilator of $K\Pi_9 e_P$ in $K\Pi_9$ for an arbitrary field K?*

Exercise 212 *Prove remark 22!*

Exercise 213 *Prove corollary 22!*

Exercise 214 *Prove lemma 10 by studying Kenneth Brown's article [7]!*

Exercise 215 *Let $n \in \mathbb{N}$. Determine an element $P \in \Pi_n$ such that 1 is no element of $M_{\geq P}$ ist. Is it possible to determine all such elements P?*

Exercise 216 *Let $n \in \mathbb{N}$. Is $\leq$ anti-symmetric on Π_n?*

Exercise 217 *Let $n \in \mathbb{N}$. Is $\leq$ symmetric on Π_n?*

Exercise 218 *Let $n \in \mathbb{N}$. Is $\leq$ connex on Π_n?*

Exercise 219 *Prove theorem 34 by studying Kenneth Brown's article [7]!*

Exercise 220 *Prove remark 23!*

Exercise 221 *Prove remark 25!*

Exercise 222 *Prove corollary 21!*

Exercise 223 *Let K be a field, $n \in \mathbb{N}$ and $P \in \Pi_n$. On what terms is ρ_P an algebra-endomorphism of $K\Pi_n$? On what terms is ρ_P injective, surjective, bijective?*

Exercise 224 *Within remark 23 we have defined the inverse relation of an arbitrary relation. Prove that this relation is symmetric resp. anti-symmetric resp. connex if and only if the original relation has this property.*

List of Tables

187

List of Figures

Bibliography

[1] Adrian Abraham Albert, Structure of algebras, Amer. Math. Soc. Colloq. Publ., vol. 24, Amer. Math. Soc, Providence, R. I., 1961. MR 23 A912.

[2] M. D. Atkinson, Solomon's descent algebra revisted, Bull. London Math. Soc. 24, 1992, 545-551

[3] Thorsten Bauer, Über die Struktur der Solomon-Algebren, Bayreuther Mathematische Schriften, Heft 63, 2001, 1-102

[4] Thorsten Bauer, Salvatore Siciliano, Carter subgroups of units of an associative algebra, Bull. Aust. Math. Soc. 71, No.3, pages 471-478, 2005

[5] H. Bell and Y. Li, Duo group rings, J. Pure Appl. Algebra, 209, pages 833-838, 2007

[6] Patrick Bidigare, Hyperplane arrangement face algebras and their associated Marcov chains, Ph.d. thesis, University of Micigan, 1997

[7] Kenneth S. Brown, Semigroup and ring theoretical methods in probability. In Representations of finite dimendional algebras and related topics in Lie theory and geometry, volume 40 of Fields Inst. Commun., pages 3-26. Amer. Math. Soc., Providence, RI, 2004

[8] R.C. Courter, Finite Dimensional Right Duo Algebras Are Duo, Proceedings of the AMS, Volume 84, Number 2, 1982

[9] Xiankun Du, The centers of a radical ring, Canad. Math. Bull. 35, no. 2, 174-179, 1992

[10] Yurij A. Drozd, Vladimir V. Kirichenko, Finite dimensional algebras, Springer-Verlag, Berlin-Heidelberg, 1994

192

[11] Bertram Huppert, Endliche Gruppen I, Springer-Verlag, Berlin-Heidelberg, 1967

[12] P.J. Morandi, B.A. Sethuramam, J.-P. Tagnol, Division algebras with an anti-automorphism but with no involution, Advances in geometry, Vol.5, Ser. 3, 2005

[13] Tadashi Nakayama, Note on Uni-serial and Generalized Uni-serial rings, Proc. Imp. Acad., Vol. 16, Number 7, 1940, pages 285-289

[14] Richard S. Pierce, Associative Algebras, Springer-Verlag, New York, 1982

[15] Wolfgang M. Ruppert, Vorlesungsskript Analytische Zahlentheorie, Wintersemester 04/05, Uni Erlangen, Part 4, www.mi.uni-erlangen.de/ ruppert/WS0405/Part4

[16] Winfried Scharlau, Automorphism and involutions of incidence algebras, Proc. Ottawa, 1994 Conf. On Repr. of algebras, Lecture Notes in Mathematics 488, Springer, Berlin, 1976, pages 340-350

[17] Günter Scheja, Uwe Storch, Lehrbuch der Algebra Part 2, BG. Teubner Stuttgart, 1988

[18] Manfred Schocker, The module structure of the Solomon-Tits algebra of the symmetric group, J. Alg. 301(2006), No.2, pages 554-586 (Peprint available at http://arxiv.org/abs/math/0505137)

[19] Salvatore Siciliano, Cartan subalgebras in Lie algebras of associative algebras, Communications in Algebra, Volume 34, Issue 12 December 2006 , pages 4513 - 4522

[20] Louis Solomon. A Mackey formula in the group ring of a Coxeter group, J. Algebra, 41, pages 255-268, 1976.

[21] Sven Wirsing, About Cartan-Subalgebras in Lie-Algebras associated to Associative Algebras, Cornell University Library, arxiv.org, 2012

[22] S. Wirsing, Separabilität in kommutativen und auflösbaren Algebren, Unter Berücksichtigung nicht-unitärer assoziativer Algebren, Disserta-Verlag, 2015, Hamburg

[23] S. Wirsing, Über Einheitengruppen modularer Gruppenalgebren, Disserta-Verlag, 2015, Hamburg

[24] S. Wirsing, Über die Struktur der Solomon-Tits-Algebren der symmetrischen Gruppe, Eine Analyse assoziativer, gruppentheoretischer und Lie-theoretischer Phänomene, Disserta-Verlag, 2015, Hamburg

[25] S. Wirsing, Maximal nilpotente Teilstrukturen I, Nilradikale und Cartan-Teilalgebren in assoziierten Lie-Algebren, Disserta-Verlag, 2015, Hamburg

[26] S. Wirsing, Maximal nilpotent substructure I, Nilradicals and Cartan subalgebras in associative algebras, Disserta-Verlag, 2016, Hamburg

[27] S. Wirsing, Maximal nilpotente Teilstrukturen II: Eine Korrespondenz in auflösbaren Algebren, Disserta-Verlag, 2016, Hamburg

[28] S. Wirsing, Maximal nilpotent substructure II: A correspondence theorem within solvable associative algebras, Anchor-Verlag, 2016, Hamburg

[29] S. Wirsing, Separability within commutative and solvable associative algebras. Under consideration of non-unitary algebras, Anchor-Verlag, 2018, Hamburg

[30] S. Wirsing, On unit groups of modular group algebras. The concept of end-commutable ordering, Anchor-Verlag, 2019, Hamburg

[31] Manfred Wolff, Peter Hauck, Wolfgang Küchlin: Mathematik für Informatik and Bio-Informatik. Springer-Verlag, Berlin Heidelberg New-York, S.50

Index